Materials Forming and Machining

Related titles

Machining: Fundamentals and Recent Advances
(ISBN 978-1-84800-212-8)

Woodhead Publishing in Mechanical Engineering

Materials Forming and Machining

Research and Development

Edited by

J. Paulo Davim

AMSTERDAM • BOSTON • CAMBRIDGE • HEIDELBERG
LONDON • NEW YORK • OXFORD • PARIS • SAN DIEGO
SAN FRANCISCO • SINGAPORE • SYDNEY • TOKYO

Woodhead Publishing is an imprint of Elsevier

WOODHEAD
PUBLISHING

Woodhead Publishing Limited is an imprint of Elsevier
80 High Street, Sawston, Cambridge, CB22 3HJ, UK
225 Wyman Street, Waltham, MA 02451, USA
Langford Lane, Kidlington, OX5 1GB, UK

ISBN: 978-0-85709-483-4 (print)
ISBN: 978-0-85709-484-1 (online)

British Library Cataloguing in Publication Data
A catalogue record for this book is available from the British Library

Library of Congress Control Number: 2015942650

For Information on all Woodhead Publishing publications
visit our website at http://store.elsevier.com/

Contents

Contributors

B. Abbaszadeh Tarbiat Modares University, Tehran, Islamic Republic of Iran

N. Bay Technical University of Denmark, Lyngby, Denmark

G. Centeno University of Seville, Seville, Spain

U.S. Dixit Indian Institute of Technology Guwahati, Guwahati, India

V.N. Gaitonde B.V.B. College of Engineering and Technology, Hubli, India

S.N. Joshi Indian Institute of Technology Guwahati, Guwahati, India

R. Kant Indian Institute of Technology Guwahati, Guwahati, India

S.R. Karnik B.V.B. College of Engineering and Technology, Hubli, India

M.M. Kasaei Tarbiat Modares University, Tehran, Islamic Republic of Iran

M. Kompitsas Theoretical and Physical Chemistry Institute (TPCI), Athens, Greece

P. Koralli National Technical University of Athens (NTUA), and Theoretical and Physical Chemistry Institute (TPCI), Athens, Greece

G. Kyriakakis National Technical University of Athens (NTUA), Athens, Greece

D.E. Manolakos National Technical University of Athens (NTUA), Athens, Greece

A.P. Markopoulos National Technical University of Athens (NTUA), Athens, Greece

A.J. Martínez-Donaire University of Seville, Seville, Spain

P.A.F. Martins Universidade de Lisboa, Lisboa, Portugal

D. Morales-Palma University of Seville, Seville, Spain

H. Moslemi Naeini Tarbiat Modares University, Tehran, Islamic Republic of Iran

J. Paulo Davim University of Aveiro, Aveiro, Portugal

I.S. Pressas National Technical University of Athens (NTUA), Athens, Greece

M.B. Silva Universidade de Lisboa, Lisboa, Portugal

C. Vallellano University of Seville, Seville, Spain

Preface

Nowadays, ways of forming materials include a large family of manufacturing processes in which plastic deformation and other techniques are used to change the shape of workpieces. Processes for forming materials include extrusion, forging, rolling, drawing, sheet metal forming, microforming, hydroforming, thermoforming, and incremental forming.

Machining is a manufacturing process in which a cutting tool or other technique is used to remove excess material from a workpiece so that the remaining material is the desired part shape. Machining includes traditional machining (turning, milling, drilling, etc.), nontraditional machining (EDM, ECM, USM, LAM, etc.), abrasive machining, hard part machining, high speed machining, high efficiency machining, and micromachining, among others.

Forming technologies and machining can be applied to a wide variety of materials, namely, metals, polymers, ceramics, composites, biomaterials, and nanomaterials.

This research book aims to provide information on materials forming and machining for modern industry. The initial chapter of the book provides novel experimental techniques for determination of the forming limits at necking and fracture. Chapter 2 is dedicated to hole-flanging by single point incremental forming. Chapter 3 presents flexible roll forming. Chapter 4 covers research issues in the laser sheet bending process. Chapter 5 is dedicated to multiple performance optimization in drilling using the Taguchi method with utility and modified utility concepts. Chapter 6 contains information on molecular dynamics simulation of material removal with the use of a laser beam. Finally, the last chapter of the book is dedicated to manufacturing processes of shape memory alloys.

This book can be used as a research book for a final undergraduate engineering course or as a topic on materials forming and machining at the postgraduate level. Also, this book can serve as a useful reference for academics; manufacturing researchers; mechanical, manufacturing, industrial, and materials engineers; and for professionals in materials forming and machining. The scientific interest in this book is evident for many important centers of research, laboratories, and universities throughout the world.

The editor acknowledges WoodHead/Elsevier for this opportunity and for their enthusiastic and professional support, and finally, I would like to thank all the chapter authors for their availability for this work.

J. Paulo Davim
University of Aveiro, Portugal
May 2015

About the contributors

Editor

J. Paulo Davim received his PhD degree in mechanical engineering from the University of Porto in 1997 and the Aggregate Title from the University of Coimbra in 2005. Currently, he is a professor at the Department of Mechanical Engineering of the University of Aveiro and the head of MACTRIB—Machining and Tribology Research Group. He has more than 28 years of teaching and research experience in manufacturing, materials, and mechanical engineering with special emphasis in Machining and Tribology. Recently he has also developed an interest in Industrial Engineering and Sustainable Manufacturing. He is the editor-in-chief of six international journals, a guest editor of several journals, and an editor of other books and series and is a scientific advisor for many international journals and conferences. Presently, he is an Editorial Board member of 25 international journals and acts as a reviewer for more than 80 prestigious ISI Web of Science journals. In addition, he has also published, as author and coauthor, more than 50 book chapters and 350 articles (more than 200 articles in ISI Web of Science, h-index 32+).

Authors

Chapter 1

Gabriel Centeno Báez received his PhD in mechanical engineering from the University of Seville (US), Spain, in 2010. He is currently an assistant professor of manufacturing engineering at the Department of Mechanical and Manufacturing Engineering of the University of Seville. His research interest is sheet metal forming with a special focus on incremental sheet forming and its applications, and he is coauthor of more than 20 papers in international journals and conferences.

Andrés Jesús Martínez-Donaire received his PhD in mechanical engineering from the University of Seville (US), Spain, in 2012. He is currently an assistant professor of manufacturing processes engineering at the Department of Mechanical and Manufacturing Engineering of the University of Seville. His research interest is sheet metal forming processes and he is coauthor of around 20 papers in international journals and conferences.

Domingo Morales Palma received his PhD in mechanical engineering from the University of Seville (US), Spain, in 2011. He is currently an assistant professor at the Department of Mechanical and Manufacturing Engineering of the University of

Seville. His research interest is sheet metal forming and he is coauthor of more than 20 papers in international journals and conferences.

Carpóforo Vallellano received his PhD in mechanical engineering from the University of Seville (US), Spain, in 1999. He is currently an associate professor of manufacturing process engineering at the Department of Mechanical and Manufacturing Engineering of the University of Seville. His research interests include metal fatigue and metal forming and he has recently focused on sheet metal forming processes. He is coauthor of more than 100 papers in international journals and conferences. He is a member of the European Structural Integrity Society (ESIS), a senior member of the Society of Manufacturing Engineers (SME), a member of the Grupo Español de Fractura (GEF), and a founding member of the Sociedad de Ingeniería de Fabricación (SIF).

Maria Beatriz Silva received her MSc in mechanical engineering from Instituto Superior Técnico (IST), University of Lisbon, Portugal, and her PhD in mechanical engineering from the same university in 2008. She is currently an assistant professor of manufacturing at IST. Her research interest is metal forming and she is coauthor of 4 book chapters and 44 papers in international journals and conferences.

Paulo A.F. Martins received his PhD in mechanical engineering from Instituto Superior Técnico (IST), University of Lisbon, Portugal in 1991 and received his Habilitation in 1999 in recognition of his work in the numerical and experimental simulation of metal forming processes. He is currently a professor of manufacturing at IST. His research interests include metal forming and metal cutting, and he is coauthor of 6 books, several national and international patents, and 300 papers in international journals and conferences. He is an associate member of CIRP (The International Academy for Production Engineering) and belongs to the Editorial Board of several international journals.

Chapter 2

Maria Beatriz Silva received her MSc in mechanical engineering from Instituto Superior Técnico (IST), University of Lisbon, Portugal, and her PhD in mechanical engineering from the same university in 2008. She is currently an assistant professor of manufacturing at IST. Her research interests are metal forming and she is coauthor of 4 book chapters and 44 papers in international journals and conferences.

Niels Bay received his MSc in mechanical engineering from Technical University of Denmark (DTU) in 1972, and his PhD from the same university in 1977. In 1987 he received the degree of Dr. Techn. (habilitation). His research interests include metal forming, metal forming tribology, cold welding, and resistance welding. He is coauthor of 1 book, 3 patents, and about 300 papers in international journals and conferences. He is former president of the ICFG (Int. Cold Forging Group) and former chairman of the Scientific Technical Committee of Forming of CIRP (The Int. Academy for Production Engineering) as well as former Danish delegate in the EU's CGC/CAN Advisory group for Industrial and Materials Technologies. He belongs to the Editorial Board of several international journals.

Paulo A.F. Martins received his PhD in mechanical engineering from Instituto Superior Técnico (IST), University of Lisbon, Portugal in 1991 and received his Habilitation in 1999 in recognition of his work in the numerical and experimental simulation of metal forming processes. He is currently a professor of manufacturing at IST. His research interests include metal forming and metal cutting, and he is coauthor of 6 books, several national and international patents, and 300 papers in international journals and conferences. He is an associate member of CIRP (The International Academy for Production Engineering) and belongs to the Editorial Board of several international journals.

Chapter 3

Mohammad Mehdi Kasaei received his MSc in mechanical engineering from Tarbiat Modares University (TMU), I.R. Iran in 2010. He is currently a PhD student in mechanical engineering at the same university. He carried out part of his PhD research work at Instituto Superior Técnico (IST), University of Lisbon, Portugal as a visiting researcher in 2014. His research interest is metal forming, and he is coauthor of 15 papers in international journals and conferences.

Hassan Moslemi Naeini received his MSc in mechanical engineering from Tarbiat Modares University (TMU), I.R. Iran in 1993 and his PhD in mechanical engineering from the Institute of Industrial Science, University of Tokyo, Japan in 2000. He is currently a professor of the Manufacturing Group at TMU. His research interest is the numerical and experimental simulation of sheet metal forming. He is coauthor of 210 papers in national and international journals and conferences. He is a member of the Japan Society on Technology of Plasticity (JSTP), the Iranian Society of Mechanical Engineers (ISME), and belongs to the Editorial Board of several national journals.

Behnam Abbaszadeh received his MSc in mechanical engineering from Tarbiat Modares University (TMU), I.R. Iran in 2014. He is currently a PhD student in mechanical engineering at the same university. His research interest is metal forming, and he is coauthor of two papers in international journals and conferences.

Maria Beatriz Silva received her MSc in mechanical engineering from Instituto Superior Técnico (IST), University of Lisbon, Portugal, and her PhD in mechanical engineering from the same university in 2008. She is currently an assistant professor of manufacturing at IST. Her research interest is metal forming, and she is coauthor of 4 book chapters and 44 papers in international journals and conferences.

Paulo A.F. Martins received his PhD in mechanical engineering from Instituto Superior Técnico (IST), University of Lisbon, Portugal in 1991 and received his Habilitation in 1999 in recognition of his work in the numerical and experimental simulation of metal forming processes. He is currently a professor of manufacturing at IST. His research interests include metal forming and metal cutting, and he is coauthor of 6 books, several national and international patents, and 300 papers in international journals and conferences. He is an associate member of CIRP (The International Academy for Production Engineering) and belongs to the Editorial Board of several international journals.

Chapter 4

Ravi Kant is a research scholar at the Department of Mechanical Engineering, IIT Guwahati, India. He is pursuing his doctoral studies in the "Laser Bending Process." In his postgraduation, he worked in the area of "Formability of Adhesive Bonded Blanks." His research interests include process modeling, optimization, and soft-computing modeling of heat transfer, forming, and machining processes. He has about 15 national and international publications.

Shrikrishna N. Joshi completed his doctoral studies in intelligent modeling and optimization of an electric discharge machining process from IIT Bombay in 2009. Since then he has worked as an assistant professor in the Department of Mechanical Engineering, IIT Guwahati. His research interests are micromachining and microbending using lasers; computer aided design and manufacturing (CAD/CAM); manufacturing process modeling and optimization; and mechatronics. He teaches undergraduate/ graduate courses on micromanufacturing; mechatronics and manufacturing automation, CAD/CAM; and manufacturing technology. He is guiding five PhD students who are working on various research areas. He has about 25 papers published in international journals and conferences of national/international repute.

Uday S. Dixit obtained a bachelor's degree in mechanical engineering from erstwhile University of Roorkee (now Indian Institute of Technology Roorkee) in 1987, an MTech in mechanical engineering from Indian Institute of Technology (IIT) Kanpur in 1993, and his PhD in mechanical engineering from IIT Kanpur in 1998. He joined the Department of Mechanical Engineering, Indian Institute of Technology Guwahati, in 1998 where he is currently a professor. He is also officiating director of Central Institute of Technology, Kokrajhar. He has been actively engaged in carrying out research in applied plasticity for the last 23 years. He has published about 55 journal papers, 50 conference papers, and 3 books related to manufacturing and the finite element method. He has also edited one book on metal forming. He has guest-edited a number of special issues of journals and is currently an associate editor of the *Journal of Institution of Engineers (India) Series C*. He has guided 5 doctoral and 37 masters students.

Chapter 5

V.N. Gaitonde is a professor in the Industrial and Production Engineering Department at B.V.B. College of Engineering and Technology, Hubli. He obtained his ME in production management from Karnataka University, Dharwad and his PhD from Kuvempu University, Shimoga. His fields of interest include process modeling and optimization, application of artificial neural networks (ANN), genetic algorithm (GA), particle swarm optimization (PSO), and robust design in manufacturing processes. He has more than 25 years of teaching and research experience. He is the associate editor for 1 international journal, an Editorial Board member of 5 international journals, a reviewer for many international journals and has published more than 70 papers in refereed international journals and conferences.

S.R. Karnik is a professor of Electrical & Electronics Engineering at B.V.B. College of Engineering and Technology, Hubli. He obtained his MTech from IIT

Kharagpur in 1993 and his PhD from VTU Belgaum in 2012. His fields of interest include process modeling and optimization, power system analysis, artificial neural networks (ANN), genetic algorithm (GA), particle swarm optimization (PSO), and robust design applications to power system monitoring and control. He has more than 27 years of teaching and research experience. He is the Editorial Board member of 4 international journals, a reviewer for many international journals, and has published more than 60 papers in refereed international journals and conferences.

J. Paulo Davim received his PhD degree in mechanical engineering from the University of Porto in 1997 and the Aggregate Title from the University of Coimbra in 2005. Currently, he is a professor at the Department of Mechanical Engineering of the University of Aveiro and the Head of MACTRIB—Machining and Tribology Research Group. He has more than 28 years of teaching and research experience in manufacturing, materials, and mechanical engineering with special emphasis in Machining and Tribology. Recently he has also developed an interest in Industrial Engineering and Sustainable Manufacturing. He is the editor-in-chief of six international journals, a guest editor of several journals, and an editor of other books and series and is a scientific advisor for many international journals and conferences. Presently, he is an Editorial Board member of 25 international journals and acts as a reviewer for more than 80 prestigious ISI Web of Science journals. In addition, he has also published, as author and coauthor, more than 50 book chapters and 350 articles (more than 200 articles in ISI Web of Science, h-index 32+).

Chapter 6

Angelos P. Markopoulos is a lecturer in the Manufacturing Technology Division at the School of Mechanical Engineering, National Technical University of Athens, Greece. His research includes topics such as precision and ultraprecision machining processes with a special interest in high speed hard machining, grinding, and micromachining. He is the author of more than 30 papers in journals, conferences, and book chapters on the above mentioned topics.

Panagiota Koralli is a PhD candidate in the Manufacturing Technology Division at the School of Mechanical Engineering, National Technical University of Athens, Greece. She is cooperating with the Laser-based Techniques and Applications Lab (LATA) of the Theoretical and Physical Chemistry Institute of the National Hellenic Research Foundation and her work concerns the growth and study of third generation photovoltaic solar cells based on the thin film technology as well as the optimization of the laser micromachining technique of thin films (laser scribing).

George Kyriakakis is a mechanical engineer. He received his master's degree from National Technical University of Athens, Greece after attending the Materials Science and Technology program in 2013.

Michael Kompitsas received his Diploma of physics (1975) and the PhD of physics (1980) from the University of Heidelberg, Germany. He joined the National Hellenic Research Foundation/Theor. and Phys. Chem. Institute in 1983 and developed the Laser Applications Laboratory. His activity includes (a) pulsed-laser deposition (PLD) and doping of semiconducting metal oxide thin films, (b) development and testing of thin film-based electrochemical sensors for toxic gases, (c) laser-induced plasma spectroscopy (LIPS) for environmental, biological, and industrial analytical work, (d) photo-physics and photo-chemistry, and (e) atomic and molecular structure.

Dimitrios E. Manolakos is a professor in the Manufacturing Technology Division at the School of Mechanical Engineering, National Technical University of Athens, Greece. He has published more than 180 papers in various areas of manufacturing technology and materials science.

Chapter 7

Angelos P. Markopoulos is a lecturer in the Manufacturing Technology Division at the School of Mechanical Engineering, National Technical University of Athens, Greece. His research includes topics such as precision and ultraprecision machining processes with a special interest in high speed hard machining, grinding and micro-machining. He is the author of more than 30 papers in journals, conferences, and book chapters on the above mentioned topics.

Ioannis S. Pressas was born in Athens in 1988. Currently, he is a PhD candidate in the School of Mechanical Engineering at the National Technical University of Athens. His field of research includes the manufacturability of advanced materials, as well as the simulation of manufacturing processes in advanced materials.

Dimitrios E. Manolakos is a professor in the Manufacturing Technology Division at the School of Mechanical Engineering, National Technical University of Athens, Greece. He has published more than 180 papers in various areas of manufacturing technology and materials science.

Novel experimental techniques for the determination of the forming limits at necking and fracture

1

G. Centeno[1], A.J. Martínez-Donaire[1], D. Morales-Palma[1], C. Vallellano[1], M.B. Silva[2], P.A.F. Martins[2]

[1]University of Seville, Seville, Spain; [2]Universidade de Lisboa, Lisboa, Portugal

1.1 Introduction

Fracture initiation and propagation in sheet metal forming is controlled by one of the following two mechanisms: (i) sheet thinning caused by localized necking along an inclined direction to the loading direction and (ii) sheet thinning caused by plastic flow without necking.

In the early 1950s, Hill (1952) associated the first fracture mechanism with plastic instability and, later in the mid 1960s, Marciniak (1965) explained the experimental observations of strain localization in hydraulic bulging and punch stretching in the light of geometrical or structural nonhomogeneity of the material. The mechanism of fracture with previous localized necking is schematically illustrated in the strain loading path OABC in Figure 1.1, which is characterized by a sharp change toward near plane strain deformation after crossing the onset of necking (also known as the forming limit curve [FLC]) at point B (Swift, 1952).

The second fracture mechanism results from the pioneering research work by Embury and Duncan (1981) in the late 1970s who showed that cracks in punch stretching can be triggered without previous localized necking. The mechanism of fracture without previous localized necking is also observed in single-point incremental forming (SPIF) of sheet metal parts (Silva, Skjoedt, Atkins, Bay, & Martins, 2008) and flanges (Centeno, Silva, Cristino, Vallellano, & Martins, 2012), and it is schematically illustrated in Figure 1.1 by means of the strain loading path ODE, which experiences no change in direction after crossing the FLC at point D.

The FLC is usually determined following the standard ISO 12004-2:2008 using Nakazima- or Marciniak-type tests, for which bending effects are a priori negligible. In the current version, the onset of necking is estimated using a position-dependent methodology, that is, analyzing the principal strain distribution on the sheet at a fixed and unique instant immediately before the crack appearance. The practical application and limitations of this method have been discussed by Hotz and Timm (2008), Martínez-Donaire, Vallellano, Morales, and García-Lomas (2009, 2010), and Martínez-Donaire, García-Lomas, and Vallellano (2014). However, the recent use of optical techniques and image analysis to online measure the strains on the sheet during testing has allowed exploring what is called time-dependent methodologies

Materials Forming and Machining. http://dx.doi.org/10.1016/B978-0-85709-483-4.00001-6

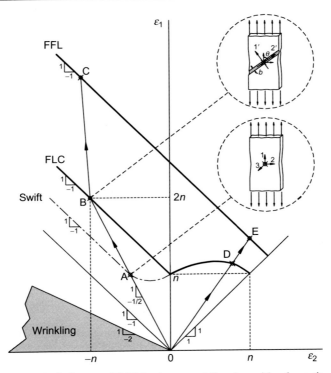

Figure 1.1 Forming limit diagram (FLD) in sheet metal forming with schematic representation of the forming limit curve (FLC), of the fracture forming limit line (FFL) corresponding to opening by tension and of the onset of diffuse necking due to Swift (1952).

for the FLC determination. In this case, the necking is estimated by analyzing the time evolution of strains at the fracture zone. A comparison of some recent time-dependent algorithms can be seen in Hotz, Merklein, Kuppert, Friebe, and Klein (2013), Martínez-Donaire, García-Lomas, & Vallellano (2014), and Wang, Carsley, He, Li, and Zhang (2014).

The first part of this chapter is focused on the description of two physically based methodologies to accurately set the onset of necking and the limit strains that were recently proposed by the authors (Martínez-Donaire, Vallellano, Morales, & García-Lomas 2009, 2010; Martínez-Donaire, García-Lomas, & Vallellano, 2014). The first one is a time-dependent methodology, and the second one is a hybrid method based on direct visualization and analysis of displacements at the outer sheet surface during the test. Both methodologies are able to detect necking not only in the conventional FLC tests, but also in situations where a significant strain gradient through the sheet thickness is expected (e.g., operations with small radii punches, corner radii in forming dies, stretch-bending operations). Most of the existing methods, designed just for Nakazima or Marciniak tests, usually fail in their predictions when they are applied to these common practical situations. A series of Nakazima and stretch-bending tests performed on aluminum alloy will give support to the presentation.

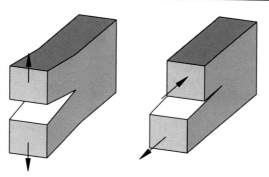

Figure 1.2 Crack opening modes in sheet metal forming: (left) mode I—tensile stresses and (right) mode II—in-plane shear stresses.

In what concerns fracture forming limits in sheet metal forming, it is important to understand first that cracking usually occurs as ductile fracture, rarely as brittle fracture, in two different opening modes: (i) tensile and (ii) in-plane shear (respectively, the same as modes I and II of fracture mechanics, Figure 1.2).

The second part of this chapter revisits forming limit diagrams (FLDs) in the light of fundamental concepts of plastic flow and ductile damage with the purpose of analyzing the circumstances under which each crack opening mode will occur. The presentation follows the analytical approach that was recently developed by Martins, Bay, Tekkaya, and Atkins (2014) for characterizing fracture loci of anisotropic metal sheets under plane-stress loading conditions. Experimental results give support to the overall methodology and confirm the existence of two different fracture loci corresponding to crack opening by tension (the fracture forming limit line [FFL]) and by in-plane shear (the shear fracture forming limit line [SFFL]).

Finally, the third and last part of this chapter presents experimental results to analyze failure of steel sheets within the FLD comprising the FLC and FFL limit curves obtained by means of the methods and procedures that will be exposed in the first two parts of the chapter. With this purpose in mind, a series of stretch-bending and SPIF tests were carried out using a variety of tool diameters in order to analyze the influence of bending on formability.

1.2 Forming limit curve

Recent publications have been proposing a number of methods to estimate the conventional FLC by means of a time-dependent analysis of the strains. Some of these methods can be found in Geiger and Merklein (2003), Situ, Jain, and Bruhis (2006, 2007), Eberle, Volk, and Hora (2008), Huang, Zhang, and Yang (2008), Feldmann, Schatz, and Aswendt (2009), Volk and Hora (2010), Merklein, Kuppert, and Geiger (2010), Sène, Balland, Arrieux, and Bouabdalah (2012), Li, Carsley, Stoughton, Hector, and Hu (2013), Hotz, Merklein, Kuppert, Friebe, and Klein (2013), and Wang, Carsley,

He, Li, and Zhang (2014), and among others. These developments highlight the current research and industrial interest in determining the onset of necking and the FLC, but none of them is now universally accepted.

The FLC provides the principal strain pairs on the sheet surface at the onset of necking. Localized necking can be physically identified as an unstable local reduction in sheet thickness. Before the onset of necking, the distribution of strain depends on the geometry of the forming tools and the geometry and material of the sheet. However, after the onset of necking, there will be a concentration of strain in the aforementioned unstable localized region with a size on the order of magnitude of the sheet thickness. As a result, the strain rate of the material placed outside the localized region of necking decreases gradually until it finally vanishes.

Figure 1.3 illustrates the behavior in the neighborhood of the necking region. It shows the time evolution at various aligned points in a section perpendicular to the fracture region. The strain level of some points (B and C) increases monotonically until fracture, while other points (A, D, and E) cease to strain and even undergo some elastic unloading immediately before failure. The first set of points is clearly located in the region of plastic instability whereas the second set of points is located in the adjacent regions. This pattern characterizes the strain localization process during the development of plastic instability and establishes the basis for the detection of the onset of necking by means of the two methods that will be described in the following sections of this chapter.

1.2.1 New time-dependent method

A new time-dependent method (t-d method) makes use of a temporal analysis of the major strain distribution and its first time derivative ($\dot{\varepsilon}_1$, "major strain rate") for a series of points on the outer sheet surface along a cross section perpendicular to the crack. Figure 1.4 is a schema of the time evolution of ε_1 and its time derivative $\dot{\varepsilon}_1$ at two representative points, A and B, in a section in the necking region. This procedure can be divided into the following steps:

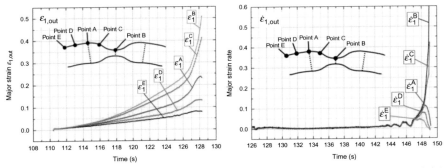

Figure 1.3 Experimental time evolution of the major strain (left) and major strain rate (right) in a stretch-bending test (⌀10 mm cylindrical punch) at various aligned points along a section perpendicular to the fracture.

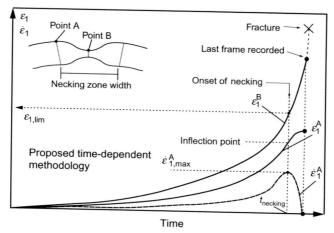

Figure 1.4 Schematic of the developed time-dependent methodology.

1. Obtaining the necking zone width. As previously discussed, the strain of any point within this region increases until fracture. Outside this region the strain rate is gradually reduced until it ceases, reaching a constant level of strain or even undergoing some elastic unloading before fracture (see Figure 1.3). The boundary of the instability region is defined by the last two points (point A in Figure 1.4) on either side of the crack that cease to strain and reach a zero strain rate just before the crack appears.

2. Determining the onset of necking ($t_{necking}$). Experimental evidence shows that the onset of necking is associated with a decreasing strain rate in the material adjacent to the necking region, vanishing before the fracture occurs. Therefore, it can be established that the necking process begins when the strain rate at the boundary of the instability region (point A in Figure 1.4) reaches a local maximum value ($\dot{\varepsilon}^A_{1,max}$). From this moment, denoted by $t_{necking}$, the strain rate at the limit of the necking zone begins to decrease to zero, signaling the start of the strain localization process inside the region.

3. Identifying the fracture point. The sheet fails at the point of greatest strain inside the necking region (point B in Figure 1.3). By definition, the strain evolution of any point in the necking region, and in particular that of point B, will always grow until failure. So the fracture point is identified as the point exhibiting the highest strain evolution curve in the necking region (ε^B_1 in Figure 1.3).

4. Determining the major limit strains at necking ($\varepsilon_{1,lim}$ and $\varepsilon_{2,lim}$). These correspond to the strains ε_1 and ε_2 at point B at the time $t_{necking}$ (see $\varepsilon_{1,lim}$ in Figure 1.3).

Similarly to the ISO 12004-2:2008, this methodology should be applied to three adjacent sections perpendicular to the crack, the final limit strain being the average value of the limit strains determined in these sections. So, for the determination of the conventional FLC, the testing conditions, such as specimen geometries, strain paths, testing directions, lubrication systems, and testing variables, should be similar to that specified in the ISO standard.

Finally, it should be noticed that the use of a main analysis variable is only motivated because this is directly measurable in the tests. Probably, from a physical perspective, the thickness strain would appear to be a more appropriate analysis variable to detect the onset of necking because it directly quantifies the reduction in the thickness. An alternative method based on a temporal analysis and its first time derivative can be implemented.

1.2.2   New time-position dependent method (flat-valley method)

This method focuses on the direct observation and analysis of the displacements of the outer surface of the test piece during the test. Figure 1.5 shows the evolution in time of the Z-displacement along a perpendicular section to the failure region in the Nakazima tests under approximate plane strain (left) and uniaxial tension (right). In both pictures a valley due to the necking process is clearly visible at the last stages of the forming process. This evidence allows formulating an alternative methodology to identify the onset of necking.

Figure 1.6 depicts a schematic of the displacement perpendicular to the plane of the nondeformed test piece (Z axis) for a section perpendicular to the fracture zone at different times (t_1 denotes a time far from necking and t_4 denotes a time just before fracture). The evolution expected can be briefly described as follows. For times far from the failure (e.g., t_1), the surface of the sheet deforms by following the curvature imposed by the tool, in our case a hemispherical or cylindrical punch. Later, this curve begins to flatten in a localized region at t_2 and becomes flat at t_3 until finally a necking valley is observed at t_4.

The outer surface begins to flatten because the thickness of the central region reduces more rapidly than at the adjacent points. When the profile becomes flat, the sheet simultaneously begins to deform locally and independently of the curvature imposed by the punch. This event physically corresponds to the onset of necking ($t_{necking} = t_3$ in Figure 1.6). At subsequent times, a valley in the sheet surface can be clearly seen, and it progressively deepens until the sheet fractures.

This flattening process can be easily identified by calculating the first spatial derivative of the Z-displacement (see Figure 1.6, bottom). So, the onset of necking corresponds graphically to a local change in the slope at the failure zone compared to

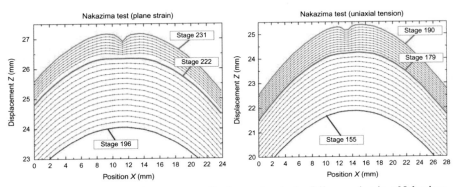

Figure 1.5 Z-displacement along a perpendicular section to the failure region in a Nakazima test under near plane strain (left) and uniaxial tension conditions (right).

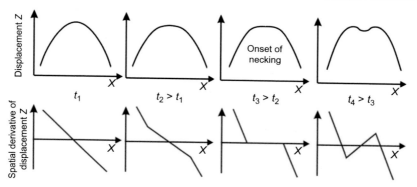

Figure 1.6 Temporal evolution of the vertical displacement profile (top) and the spatial derivative (bottom) in a section perpendicular to the failure region.

the slope of the neighboring points. Plastic instability begins when this slope remains constant, that is, when the curve becomes flat. As before, once the onset of necking is identified, the limit strains can be obtained at their corresponding strain curve at the fracture point at a time equal to $t_{necking}$.

It should be noticed that in the case of Marciniak tests, where the slope is constant from the beginning, the onset of necking should be associated with the last stage in which the slope is no longer constant, that is, the last stage before a valley is visible.

This method, referred to in the literature as the "flat-valley method," can be classified as a hybrid method that depends on both the time and the position evolution of displacements at the outer surface of the sheet around the fracture zone, that is, a time-position dependent method.

A similar approach has been very recently proposed by Wang, Carsley, He, Li, and Zhang (2014) for FLC determination based on monitoring changes in the height of the surface topography between a point within the neck to a point within the area of uniform deformation. This corroborates the feasibility of methods based on the direct observation of the surface geometry to detect the onset of localized necking.

Finally, it should be noticed that both methodologies presented above are local, that is, they focus on the evaluation of variables (strains, displacements, etc.) near the failure zone, no matter what the specimen geometry or testing conditions. Therefore, they can be applied not only to near pure stretching conditions (e.g., Nakazima or Marciniak tests), but also to situations in which there is a nonnegligible strain gradient through the thickness or along the sheet surface (e.g., in stretching processes with small-radius punches and stamping processes with high-curvature dies).

1.3 Fracture forming limits

There are three main reasons that motivated the authors to include the formability limits by fracture in this chapter. Firstly, the acceptance that engineers and technicians currently involved in the design of sheet metal parts prefer to apply design guidelines based on the critical thickness reduction than on the FLCs. Secondly, the well-known evidence that standard procedures for determining FLCs despite their simplicity and wide usage can fall short in the determination of the onset of necking due to difficulties in measurements. This often leads to the fact that FLCs of the same material provided by different sources may be different from each other. Thirdly, the understanding that currently available finite element computer programs that make use of ductile damage modeling for predicting the onset of failure require determination of the critical values of damage at the onset of fracture.

This section is focused on the connection between the strains and stresses derived from plane-stress anisotropic plasticity and ductile damage mechanics for two different crack opening modes (opening by tensile stresses—mode I, and by in-plane shear stresses—mode II).

1.3.1 Mode I—Tensile fracture

Experiments show that, irrespective of the initial loading history before necking in sheet metal forming, tensile fracture occurs approximately at a constant through-thickness true strain $\varepsilon_{3f} = \text{Const}$. Constant ε_{3f} corresponds to a constant percentage of reduction in thickness at fracture R_f given by $(t_0 - t_f)/t_0$, where t_0 is the initial thickness of the sheet and t_f is the thickness at fracture. R_f and ε_{3f} are related by $\varepsilon_{3f} = \ln(1 - R_f)$.

Owing to constancy of volume $\varepsilon_{1f} + \varepsilon_{2f} + \varepsilon_{3f} = 0$ during plastic flow, it follows that

$$\varepsilon_{1f} = -\varepsilon_{2f} - \varepsilon_{3f} \tag{1.1}$$

So, the locus of tensile fracture (designated as FFL in Figure 1.1) in the principal strain space is a straight line falling from left to right with slope "−1" (refer also to Figure 1.7, where lines of constant R_f are schematically plotted).

Figure 1.7 also presents a schematic evolution of the Mohr circles of strain for two proportional loading paths (0C and 0F) corresponding to uniaxial tension and equal biaxial stretching. For the purpose of simplifying the presentation, both loading paths are taken as linear up to the onset of fracture, without experiencing the change in direction toward plane strain conditions that one would expected to occur after crossing the FLC (not shown in the figure). This simplification may be justified on the basis of the experiments by Isik, Silva, Tekkaya, and Martins (2014) who showed that nonlinear loading paths derived from sheet formability tests in which necking preceded cracking, and linear strain paths derived from SPIF tests performed with conical and pyramidal geometries where necking was absent, provide the same fracture locus (FFL).

As shown in Figure 1.7, in the case of uniaxial tension, the diameters of the Mohr's strain circles increase as deformation progresses from A to fracture at point C but this increase is not concentric due to the necessity of ensuring the ratios between the principal strains $(\varepsilon_1 : \varepsilon_2 : \varepsilon_3 = 1 : -0.5 : -0.5)$. In the case of equal biaxial stretching, the diameter of the circles increases as deformation progresses from D to fracture at point F, but this increase is not concentric due to the necessity of ensuring the ratios

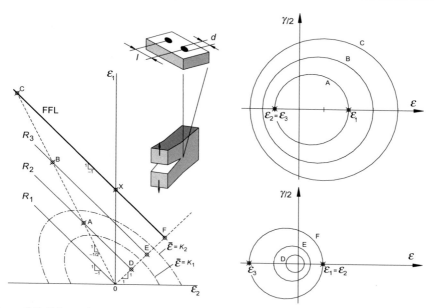

Figure 1.7 Schematic representation of the FFL in the principal strain space together with Mohr circles of strains and strain loading paths corresponding to different points A and B before fracture at C and D and E before fracture at F.

between the principal strains $\left(\varepsilon_1 : \varepsilon_2 : \varepsilon_3 = 0.5 : 0.5 : -1.0\right)$. The overall levels of strain are smaller than in the case of uniaxial tension.

Fracture is sometimes said to take place when the effective strain reaches a critical constant value, so also shown in Figure 1.7 are ellipses of constant effective strain $\bar{\varepsilon} = K_1$ and $\bar{\varepsilon} = K_2$ to indicate where failure strain pairs should lie, were cracking to occur at a critical constant value of effective strain equal to K_1 or K_2.

Because experiments show that a criterion of constant effective strain at fracture is incorrect, it is widely accepted to perform its modification by means of a weighting function that is capable of bringing other factors linked to the topic of damage mechanics into account. In practical terms this leads to the following modified version of the effective strain fracture criterion:

$$D_{\text{crit}} = \int_0^{\bar{\varepsilon}_f} g \, d\bar{\varepsilon} \tag{1.2}$$

where the nondimensional term g is a weighting function that corrects the accumulated value of the effective strain until fracture $\bar{\varepsilon}_f$ as a function of the strain loading paths.

The concept of weighting function was proposed by Bao and Wierzbicki (2004) and originally referred to a function of the components or invariants of the stress tensor. Other authors, for example Li, Luo, Gerlach, and Wierzbicki (2010), utilized a weighting function built upon stress triaxiality and lode angle parameters to formulate a modified Mohr–Coulomb fracture criterion.

However, when the weighting function is taken as the normalized stress triaxiality $g = \sigma_m / \bar{\sigma}$, Equation (1.2) results in the noncoupled void growth damage-based criterion that is directly related to the original work of McClintock (1968),

$$D_{crit} = \int_0^{\bar{\varepsilon}_f} \frac{\sigma_m}{\bar{\sigma}} d\bar{\varepsilon} \tag{1.3}$$

This is because the relation between D_{crit} and the microstructural void parameters related to the inter-hole l (interparticle inclusion) spacing and average diameter d of the hole (particle) can be formulated as $\int (\sigma_m / \bar{\sigma}) d\bar{\varepsilon} \propto \ln(l/d)$ (refer to nomenclature in Figure 1.7) (Atkins, 1996).

The integrand $\sigma_m / \bar{\sigma}$ and the variable of integration $d\bar{\varepsilon}$ in Equation (1.3) may be expressed as the product of three partial ratios involving the increment of strain $d\varepsilon_1$ and the stress σ_1. Then, using the constitutive equations associated with Hill's 48 anisotropic yield criterion (Hill, 1948), and assuming rotational symmetry anisotropy $r_\alpha = r = \bar{r}$, where $\bar{r}$ is the normal anisotropy, it is possible to rewrite Equation (1.3) as follows:

$$D_{crit} = \int_0^{\varepsilon_{1f}} \frac{\sigma_m}{\sigma_1} \frac{\sigma_1}{\bar{\sigma}} \frac{d\bar{\varepsilon}}{d\varepsilon_1} d\varepsilon_1 = \int_0^{\varepsilon_{1f}} \frac{(1+r)}{3}(1+\beta)d\varepsilon_1 \tag{1.4}$$

where $\beta = d\varepsilon_2 / d\varepsilon_1$ is the slope of a general proportional strain path and the three partial ratios σ_m/σ_1, $\sigma_1/\bar{\sigma}$, and $d\bar{\varepsilon}/d\varepsilon_1$ are given by

$$\frac{\sigma_m}{\sigma_1} = \frac{1+\alpha}{3} = \frac{(1+2r)(1+\beta)}{3[(1+r)+r\beta]}, \quad \frac{\sigma_1}{\bar{\sigma}} = \frac{1}{\sqrt{1+2r}} \frac{[(1+r)+r\beta]}{\sqrt{1+\frac{2r}{1+r}\beta+\beta^2}}, \tag{1.5}$$

$$\frac{d\bar{\varepsilon}}{d\varepsilon_1} = \frac{1+r}{\sqrt{1+2r}} \sqrt{1+\frac{2r}{1+r}\beta+\beta^2}$$

where $\alpha = \sigma_2 / \sigma_1$. The algebraic procedure that was utilized for obtaining the identity corresponding to $\sigma_m / \bar{\sigma}$ and for building up the two other partial ratios $\sigma_1 / \bar{\sigma}$ and $d\bar{\varepsilon} / d\varepsilon_1$ of Equation (1.4) is comprehensively described by Martins, Bay, Tekkaya, and Atkins (2014).

The integrand in Equation (1.4) has the form $(A + B\beta)$, implying that the damage function for a constant strain ratio β, is independent of the loading path history. Path-independence of damage functions based on integrand terms in the form $(A + B\beta)$, where A and B are constants is comprehensively discussed in Atkins and Mai (1985) and provides additional justification for the reason why strain loading paths in Figure 1.7 were assumed as linear.

Rewriting Equation (1.4) as a function of the major and minor in-plane strains $(\varepsilon_{1f}, \varepsilon_{2f})$ at the onset of fracture,

$$D_{crit} = \frac{(1+r)}{3}(\varepsilon_{1f} + \varepsilon_{2f}) \tag{1.6}$$

it follows that the critical value of the stress triaxiality $\sigma_m / \bar{\sigma}$ based damage criterion (1.3) is also a straight line with slope "-1" falling from left to right in agreement with the FFL and the condition of critical thickness reduction.

If the lower limit of the integral in Equation (1.4) is $\bar{\varepsilon}_0$ rather than zero, corresponding to situations where there is a threshold strain $\bar{\varepsilon}_0$ below which damage is not accumulated, Equation (1.6) becomes

$$D_{\text{crit}} = \frac{(1+r)}{3}\left[\varepsilon_{1f} + \varepsilon_{2f} - (1+\beta)\varepsilon_0\right] \tag{1.7}$$

meaning that the slope "−1" is unaffected, but the intercept "X" in Figure 1.7 is a function of the strain ratio β. The "upward curvature" tail of the FFL was originally discussed by Atkins (1997) and is schematically plotted in Figure 1.8.

Muscat-Fenech, Arndt, and Atkins (1996) made the connection between the FFL and fracture toughness in mode I. Taking their observation in conjunction with the above conclusions regarding the dual condition of the critical thickness reduction R_f and the critical ductile damage D_{crit} being constant and independent from the material deformation history up to fracture, it is possible to conclude that the FFL is a material property in contrast to the FLC that is not a pure material property independent of strain path.

1.3.2   Mode II—In-plane shear fracture

From a micro-based damage mechanics point of view the problem of characterizing fracture by shear is difficult because damage models based on spherical void growth such as the Gurson–Tvergaard–Needleman model (Gurson, 1977; Tvergaard & Needleman, 1984) predict no increase in damage and, therefore, no failure by fracture in situations of zero or negative stress triaxiality $\sigma_m / \bar{\sigma}$. A recent solution for this problem gave rise to a modification of the Gurson–Tvergaard–Needleman model to include a shear term that is related to damage growth in pure shear (Nahshon & Hutchinson, 2008).

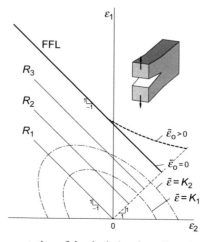

Figure 1.8 Schematic representation of the deviation from linearity of the FFL.

The formulation of the analytical conditions under which fracture occurs by in-plane shear stresses will be informed by Mohr's circle of strains, from which it may be concluded that straight lines γ_1, γ_2, and γ_3 rising from left to right and corresponding to maximum values of the in-plane distortion γ_{12} have slope "+1" and are perpendicular to the FFL (Figure 1.9).

In-plane distortions γ_{12} (hereafter designated as "γ") are caused by in-plane shear stresses τ_{12} (hereafter designated as "τ") and, therefore, it is likely that the in-plane shear fracture limiting locus (SFFL) will coincide with a straight line of slope equal to "+1," in which the major and minor in-plane strains and distortions take critical values at fracture

$$\varepsilon_{1f} - \varepsilon_{2f} = \gamma_f \tag{1.8}$$

where $\gamma_f = Y$ (Figure 1.9).

Figure 1.9 also presents a schematic evolution of the Mohr circles of strain for a loading path OAB consisting of pure shear. As shown, the diameters of the circles increase concentrically as deformation progresses due to the necessity of ensuring the ratios between the principal strains $\left(\varepsilon_1 : \varepsilon_2 : \varepsilon_3 = 1 : -1 : 0 \right)$. Fracture occurs at point B in opening mode II.

Damage mechanics also predicts a slope "+1" for the SFFL when the weighting function g in Equation (1.2) is taken as the in-plane shear stress ratio $g = \tau / \bar{\sigma}$,

$$D^s_{crit} = \int_0^{\bar{\varepsilon}_f} \frac{\tau}{\bar{\sigma}} \, d\bar{\varepsilon} \tag{1.9}$$

The weighting function g in Equation (1.9) may be thought of as a correction of the accumulated values of the effective strain until fracture at $\bar{\varepsilon}_f$ by means of the maximum in-plane shear stress τ corresponding to different strain loading paths. In

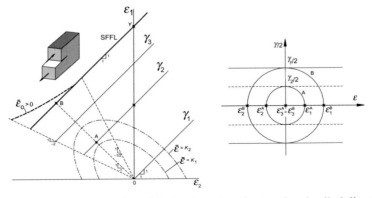

Figure 1.9 Schematic representation of the in-plane shear fracture forming limit line (SFFL) in the principal strain space together with Mohr circles of strain and a strain loading path corresponding to point A before fracture at B. The dashed tail of the SFFL represents the deviation from linearity.

other words, the new damage function combines the old fracture criterion of maximum shear stress and the equivalent plastic strain at fracture.

By expressing the integrand $\tau / \bar{\sigma}$ and the variable of integration $d\bar{\varepsilon}$ as the product of two partial ratios involving the increment of strain $d\varepsilon_1$, the in-plane shear stress τ and using the constitutive equations associated with Hill's 48 anisotropic yield criterion (Hill, 1948), it is possible to rewrite Equation (1.9) as follows:

$$D_{crit}^s = \int_0^{\bar{\varepsilon}_f} \frac{\tau}{\bar{\sigma}} d\bar{\varepsilon} = \int_0^{\varepsilon_{1f}} \frac{\tau}{\bar{\sigma}} \frac{d\bar{\varepsilon}}{d\varepsilon_1} d\varepsilon_1 = \int_0^{\varepsilon_{1f}} \frac{1}{2} \frac{(1+r)}{(1+2r)} (1-\beta) d\varepsilon_1 \qquad (1.10)$$

where the ratio $d\bar{\varepsilon} / d\varepsilon_1$ was previously defined in (1.5) and the in-plane shear stress τ and the shear stress ratio $\tau / \bar{\sigma}$ are given by

$$\tau = \frac{\sigma_1 - \sigma_2}{2} = \frac{1-\alpha}{2} \sigma_1, \quad \frac{\tau}{\bar{\sigma}} = \frac{1}{2} \frac{1}{\sqrt{1+2r}} \frac{1-\beta}{\sqrt{1+\frac{2r}{1+r}\beta + \beta^2}} \qquad (1.11)$$

When linear strain paths β in conjunction with the integrand form $(A + B\beta)$ are employed, the critical value of damage D_{crit}^s in Equation (1.10) is strain path independent and may be expressed as a function of the major and minor in-plane strains at the limiting locus of in-plane shear fracture,

$$D_{crit}^s = \frac{1}{2} \frac{(1+r)}{(1+2r)} (\varepsilon_{1f} - \varepsilon_{2f}) \qquad (1.12)$$

This result shows that critical values of damage by in-plane shear are located along a straight line rising from left to right with a slope equal to "+1" in agreement with the condition of critical distortion γ_f along the SFFL.

By following a procedure similar to that performed in case of the FFL, if the lower limit of the integral in Equation (1.10) is $\bar{\varepsilon}_0$ rather than zero, corresponding to situations where there is a threshold strain $\bar{\varepsilon}_0$ below which damage is not accumulated, Equation (1.12) becomes

$$D_{crit}^s = \frac{1}{2} \frac{(1+r)}{(1+2r)} \left[\varepsilon_{1f} - \varepsilon_{2f} - (1-\beta)\varepsilon_0 \right] \qquad (1.13)$$

meaning that the slope "+1" is unaffected, but the intercept "Y" in Figure 1.9 is a function of the strain ratio β. This leads to an "upward curvature" tail of the SFFL as it is schematically plotted by a dashed solid line in Figure 1.9.

1.4 Results and discussion

This section discusses the practical application of the methodologies previously described for the prediction of both the onset of necking (FLC) and fracture (FFL/SFFL). The importance of these two limiting curves in stretch-bending and SPIF processes are also analyzed, pointing out the influence of bending on formability. Results are comprehensively described and illustrated by a series of examples from experimental testing on steel and aluminum alloys.

1.4.1 Forming limit curves

The ability of the previously described methodologies for necking detection is discussed using an aluminum alloy 7075-O sheet with 1.6 mm thickness. A series of Nakazima tests (⌀100 mm diameter punch) are conducted according the standard ISO 12004-2:2008. On the other hand, a number of stretch-bending tests close to plane strain conditions using cylindrical punches with different radii, ⌀20, 10, and 5 mm, are also presented in order to discuss the bending effect due to the punch curvature. The strain history over the outer surface during the tests is evaluated via the digital image correlation (DIC) technique using ARAMIS®. A recording rate of 10 frames per second is set. A detailed description of the test procedure can be found in Martínez-Donaire, García-Lomas, and Vallellano (2014).

Figure 1.10 (left) shows the major strain profile along a section perpendicular to the failure region at several stages for a Nakazima (top) and a stretch-bending (bottom) test close to plane strain conditions. Such a section is shown on the major strain contour map at the frame immediately before the specimen fracture. Figure 1.10 (right) depicts the temporal evolution of major strain for a series of points distributed from the fracture outward along the previous section for the same tests.

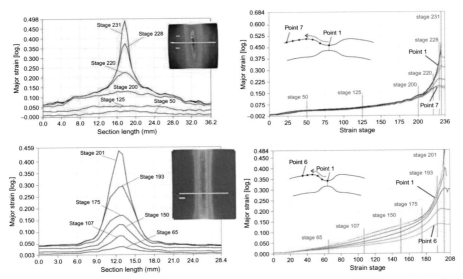

Figure 1.10 Distribution of the major true strain along a given section line (left) and evolution of the major true strain versus the strain stage at different points (right) for a Nakazima test (top) and a stretch-bending test using Φ20 mm cylindrical punch (bottom) near plane strain.

As can be seen in Figure 1.10 (left), in the early stages of the Nakazima test the strain is distributed approximately uniformly over the entire contact region between the punch and the sheet, whereas for the stretch-bending test the strain is highly localized in the central region, exhibiting a dome-shaped spatial strain profile. Accordingly, the temporal evolution of the major strain curves (see Figure 1.10 (right)) for the latter one are substantially separated from the beginning of the test, unlike in the Nakazima test, because of the severe strain gradient imposed by the punch along the surface and across the thickness of the sheet. However, in the last stages for both the Nakazima and stretch-bending tests, the curves dramatically separated from each other as a consequence of the development of the necking process. These evolutions are consistent with those shown schematically in Figure 1.4.

Applying the t-d method and the flat-valley method described above, the FLC is obtained. Figure 1.11 (left) compares both predictions with the FLC via the ISO standard. As can be seen, the results of the two developed methods are practically identical for all of the strain paths and very similar to those estimated by ISO 12004-2:2008. The maximum differences between the three approaches are in the range of 5–7% (Martínez-Donaire, García-Lomas, & Vallellano, 2014).

Figure 1.11 (right) plots the necking major strain predicted in the stretch-bending tests versus the t_0/R ratio, which quantifies the severity of the bending. It should be reminded that, due to the severe through-thickness strain gradient induced in these tests, the current ISO standard cannot be applied (Martínez-Donaire, Vallellano, Morales, & García-Lomas, 2010; Martínez-Donaire, García-Lomas, & Vallellano, 2014). As can be observed, the t-d and flat-valley methods adequately reproduce the expected experimental trend, that is, the higher the bending severity the higher the limit strains (Atzema, Fictorie, van den Boogaard, & Droog, 2010; Morales, Martínez-Donaire, Vallellano, & García-Lomas, 2009; Morales-Palma, Vallellano, & García-Lomas, 2013; Tharret & Stoughton, 2003). However, the most noteworthy result is that the predictions of both methods agree reasonably with each other, showing a small dispersion. The maximum deviations oscillated approximately 5–10% around the average values (Martínez-Donaire, García-Lomas, & Vallellano, 2014).

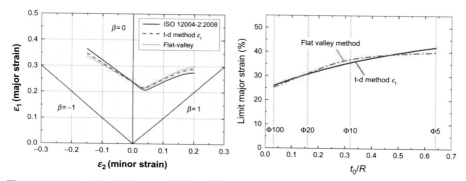

Figure 1.11 Forming limit curve (left) and limit major strain evolution versus t_0/R for tests near plane strain (right) in AA7075-O using several experimental methodologies.

In summary, the two physically based methodologies (t-d and flat valley) successfully estimate the onset of necking and the limit strains in sheet forming under significant through-thickness strain gradients and for cases with negligible bending effects. As mentioned before, the local character of these methods enables their application independent of the type of test performed, Nakazima tests, stretch-bending tests, etc.

1.4.2 Fracture loci

Instead of determining the experimental values of strains at fracture by means of conventional sheet formability tests that are commonly applied for the determination of the FLCs, authors applied a new procedure that combines results from SPIF tests, double-notched tensile tests, plane shear tests, and torsion shear tests.

The main reasons for not making use of conventional sheet formability tests are the following:

(i) Firstly, the onset of necking, implying transition from the FLC toward the FFL, is characterized by a rapid time-based bend of the strain paths toward the vertical direction due to localization of plastic instability.

(ii) Secondly, the majority of the sheet formability tests combine bending with in-plane loading over the entire surface of the specimens. Apart from the tensile, plane, and torsion shear tests, rarely are other testing operating conditions (performed with or without limited influence of bending) employed.

(iii) Thirdly, the FFL cannot be directly obtained from in-plane strain measurements performed on conventional sheet formability tests. In case of circle grid analysis, for example, the application of grids even with very small circles in order to obtain the in-plane strains in the necked region after it forms and, therefore, close to fracture, will always provide values that cannot be considered the fracture strains. This is because grids with very small circles are difficult to measure and values are dependent on the initial size of the circles in the neighborhood of the cracks due to inhomogeneous plastic deformation.

Recently published work by Isik, Silva, Tekkaya, and Martins (2014) showed that truncated conical and pyramidal SPIF tests could be successfully employed to determine the tensile fracture locus (FFL—mode I) in aluminum alloy AA1050-H111 sheets with 1 mm thickness and to avoid the first two above listed drawbacks of conventional sheet formability tests. This is because SPIF tests based on truncated conical and pyramidal geometries with varying drawing angles ψ are capable of ensuring proportional (linear) strain paths from beginning until fracture with limited and localized bending effects.

The third drawback related to in-plane strain measurement justifies the reason why the experimental values of the critical strains along the formability limits by fracture required authors to measure the thicknesses before and after fracture at different locations along the crack in order to obtain the "gauge length" strains (Atkins, 1996).

The resulting FFL is shown in Figure 1.12 where additional fracture strain points determined from double-notched tensile test specimens were added on top of the results that were previously determined by Isik, Silva, Tekkaya, and Martins (2014). The new fracture strain pairs determined from double-notched tensile tests not only show compatibility with previous values determined by SPIF but also prove the relation between the FFL and fracture by opening mode I.

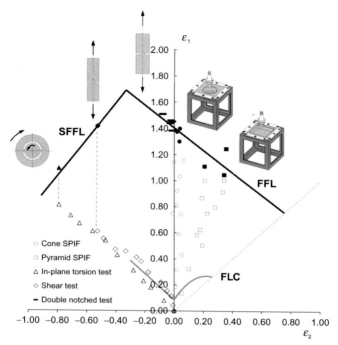

Figure 1.12 Determination of the fracture loci (FFL and SFFL) from the experimental strains at fracture that were obtained from the SPIF, double-notched tensile, and shear tests. The solid black markers correspond to strains at fracture. The inset pictures show details of the cracks for the different types of test specimens.

The location of the in-plane shear fracture locus (SFFL) in principal strain space was determined by Isik, Silva, Tekkaya, and Martins (2014) by means of torsion and in-plane shear tests but, in strict theoretical terms and discharging the development of mixed-crack separation modes at the transition between tensile and in-plane shear fracture modes, the SFFL only requires the experimental values of one type of shear test (for instance, the torsion shear test) due to the perpendicular orientation to the FFL. However, good experimental practice recommends the utilization of both torsion and in-plane shear tests instead of assuming the perpendicularity condition.

The interpolation of the fracture strain pairs given in Figure 1.12 provide the following results for the FFL and SFFL of aluminum AA1050-H111 sheets with 1 mm thickness,

$$\varepsilon_{1f} + 0.86\varepsilon_{2f} = 1.40 \quad (\text{FFL})$$
$$\varepsilon_{1f} - 1.39\varepsilon_{2f} = 2.14 \quad (\text{SFFL})$$

(1.14)

The slopes of the interpolated lines are in fair agreement with the theoretical slopes equal to "−1" and "+1" that were predicted by Equations (1.6) and (1.12) but the resulting angle between them (~85°) is in excellent agreement with the condition of perpendicularity between the two fracture lines, under the assumption of no development of mixed-crack separation modes.

The fact that the slopes of the FFL and the SFFL are perpendicular, yet they are different from "+1" and "−1" suggests that there is a missing link in the proposed analytical framework that ought to make the slopes dependent on some other effects that are not included in the approach such as, coupled ductile damage, the existence of a threshold strain $\bar{\varepsilon}_0$ below which damage is not accumulated, and the utilization of other yield criteria than Hill's 48 (Hill, 1948) that are more appropriate for modeling plastic flow of aluminum alloys, among others. This requires future research work.

1.4.3   Failure loci in single-point incremental forming and stretch-bending

As a consequence of the previous framework to determine formability limits at necking (FLC) and fracture (FFL), stretch-bending and SPIF were analyzed within a FLD obtained from Nakazima tests in AISI 304 sheets with 0.8 mm thickness. These tests were carried out with three different blank geometries (see Figure 1.13) and the overall methods and procedures, namely DIC techniques and circle grid analysis, are explained in previous Sections 1.2.1 and 1.2.2. In order to guarantee similar bending

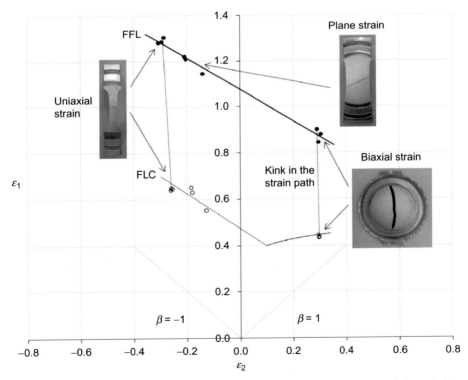

Figure 1.13 FLD based on conventional Nakazima tests containing the FLC and the FFL for AISI 304 sheets with 0.8 mm thickness.

conditions in both stretch-bending and SPIF tests, the diameter of the punches and the forming tools were made identical.

In contrast, the FFL was determined by means of the experimental methodology that was explained in previous Section 1.2.2. The FFL is approximated by a straight line $\varepsilon_1 + 0.69\varepsilon_2 = 1.08$ falling from left to right with a slope close to "−1" as originally proposed by Atkins (1997). The changes in the strain paths toward the plane strain direction after crossing the FLC are almost vertical for all test cases apart from slight deviations to the left as a result of rotation of principal directions, according to Atkins (1996).

The experimental plan in stretch-bending and SPIF is comprehensively described in Centeno, Bagudanch, Martínez-Donaire, García-Romeu, and Vallellano (2014). Due to the tool diameters of ∅20 and ∅10 mm used in both processes, a significant through-thickness strain gradient was induced due to the bending introduced by the curvature of the punches and forming tools. In addition, and contrary to what is utilized in typical incremental forming experimental apparatus, the ratio $\Phi_{backing\ plate}/\Phi_{tool}$ between the diameter of the backing plate (75 mm) and the diameter of the forming tool can be small enough to promote the development of necking in some of the experiments (Silva, Nielsen, Bay, & Martins, 2011).

Figure 1.14 shows cross sections of stretch-bending and SPIF parts produced with punches and forming tools of 20 mm diameter in conjunction with the corresponding strain paths within the FLD. As seen in Figure 1.14 (top left), the cross section in stretch-bending allows visualizing the onset of necking, whereas the cross section in SPIF, besides showing the expected reduction in thickness due to incremental forming, also allows identifying a neck (see Figure 1.14, top right). Moreover, the results of the strain paths shown in Figure 1.14 (bottom) reveal that SPIF undergoes stable plastic deformation well above the FLC, and that stretch-bending experiences failure by necking for a level of strains just slightly above the FLC. The necking strains were directly measured with a microscope (average necking strains represented by cross-shaped marks) using the sheet thickness just delimiting the unstable region, as shown in the scheme placed up-leftwards in Figure 1.14 (bottom). Comparing the strain paths in stretch-bending and SPIF, using similar punch and tool diameters, there is a clear enhancement of formability well above the FLC in the case of the latter. In fact, the parameter Δ, defined as the percentage enhancement of formability over the FLC (represented in Figure 1.14 (top left) for stretch-bending and SPIF), reaches values around 30% in stretch-bending (Δ_{S-B}) and slightly above 100% in SPIF (Δ_{SPIF}).

Figure 1.15 shows the cross sections of stretch-bending and SPIF parts produced with punches and forming tools of 10 mm diameter in conjunction with the corresponding strain paths in the FLD.

The results in Figure 1.15 (top left) show that failure in stretch-bending occurs once again by incipient necking prior to fracture for a level of strain slightly above the FLC. In contrast, the cross section of the SPIF part does not show a neck (see Figure 1.15, top right). In fact, a series of grooves can be seen on the outer surface of this part, which may be related to transition to fracture in the absence of necking. In this sense, Figure 1.15 (bottom) shows that the increase of formability in SPIF is much higher than in stretch-bending. Values of Δ_{S-B} around 30% are also obtained and values of Δ_{SPIF} are now raised up to 150%. This failure mode corresponds to the transition from failure by necking to

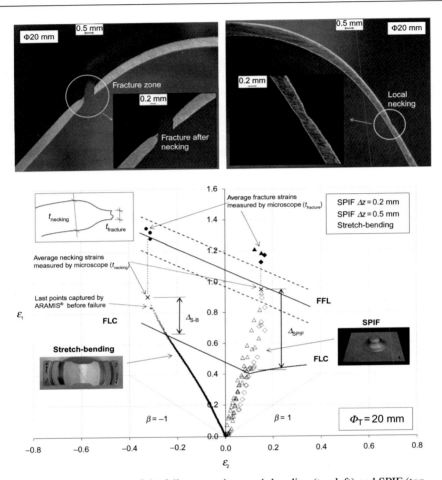

Figure 1.14 Cross sections of the failure zone in stretch-bending (top left) and SPIF (top right). Strain paths in stretch-bending and SPIF for a tool diameter of 20 mm (bottom).

failure by ductile fracture, which is commonly seen in sheet metal parts produced by SPIF with high t_0/R ratios. This phenomenon has been identified in several recent research publications such as Centeno, Silva, Cristino, Vallellano, & Martins (2012), Silva, Skjoedt, Atkins, Bay, and Martins (2008), and Silva, Skjoedt, Bay, and Martins (2009) and is comprehensively described in Silva, Nielsen, Bay, and Martins (2011).

To conclude, the main failure mode observed for the AISI 304 sheets was postponed necking followed by ductile fracture, evolving toward failure by fracture in the absence of necking in SPIF parts produced with the 10 mm diameter forming tool. Comparing the influence of bending in both stretch-bending and SPIF, it was observed that the enhancement of formability above the FLC in SPIF is much higher than in stretch-bending for the same value of t_0/R. In fact, the percentage of enhancement of formability remains just around 30% in stretch-bending for the two diameters that

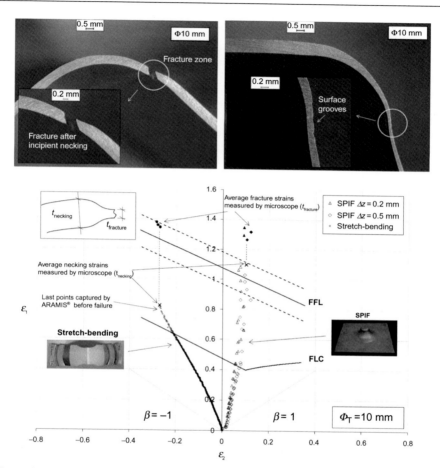

Figure 1.15 Cross sections of the failure zone in stretch-bending (top left) and SPIF (top right). Strain paths in stretch-bending and SPIF for a tool diameter of 10 mm (bottom).

were analyzed, whereas the enhancement of formability reached values up to 150% in the case of the 10 mm diameter forming tool. This means that, despite bending induced by the punch, radius plays an important role in SPIF, it cannot be considered the unique factor promoting the development of stable plastic deformations for strain values well above the FLC.

1.5 Conclusions

This chapter presented a series of innovative methods and procedures that were recently developed for the determination of the formability limits by necking and fracture. The application of these formability limits to processes involving bending, such as stretch-bending and SPIF was comprehensively discussed and the following main conclusions can be outlined:

- The proposed methods for necking detection (time-dependent and flat valley) match satisfactorily the results given by ISO 12004-2:2008 standard, becoming an appealing alternative for the determination of the FLCs. In opposition to ISO standard, these two developed methods are also able to estimate successfully the onset of necking and limit strains in sheet forming under significant through-thickness strain gradients, which are typical in industrial applications. The local character of these methods enables its application independently of the type of test performed.
- A new vision for the formability limits at fracture in sheet metal forming has also been presented. The vision makes use of fundamental plastic flow concepts related to the critical thickness reduction and the critical distortion in sheet metal forming. Fracture loci were characterized by means of a modified version of the effective strain fracture criterion that corrects the values of accumulated damage and, in contrast to alternative methods available in the literature, the proposed approach is exclusively based in sheet metal tests due to the plane-stress plastic flow conditions that are typical of sheet metal forming processes.
- The influence of bending in stretch-bending and SPIF has been analyzed within the FLD obtained using the proposed methodologies. It has been observed that the enhancement of formability above the FLC in SPIF is much higher than in stretch-bending for the same value of the bending ratio t_0/R. This means that, despite bending induced by the punch, radius plays an important role in the enhancement of formability in SPIF; it cannot be considered the unique factor promoting the development of stable plastic deformations for strain values well above the FLC.

Acknowledgments

The authors wish to thank the Spanish Government for its financial support through the research project DPI2012-32913. The authors would also like to thank the *Grup de Recerca en Enginyeria de Producte, Procés i Producció* (GREP) of the University of Girona for the support provided in AISI 304 SPIF tests.

References

Atkins, A. G. (1996). Fracture in forming. *Journal of Materials Processing Technology, 56*, 609–618.
Atkins, A. G. (1997). Fracture mechanics and metalforming: Damage mechanics and the local approach of yesterday and today. In H. P. Rossmanith (Ed.), *Fracture research in retrospect* (pp. 327–350). Rotterdam: AA Balkema.
Atkins, A. G., & Mai, Y. W. (1985). *Elastic & plastic fracture*. Chichester: Ellis Horwood.
Atzema, E. H., Fictorie, E., van den Boogaard, A. H., & Droog, J. M. M. (2010). The influence of curvature on FLC's of mild steel, (A)HSS and aluminium. In *Presented at the International Deep Drawing Research Group IDDRG, Graz, Austria* (pp. 519–528).
Bao, Y., & Wierzbicki, T. (2004). A comparative study on various ductile crack formation criteria. *Journal of Engineering Materials and Technology, Transactions of the ASME, 126*, 314–324.
Centeno, G., Bagudanch, I., Martínez-Donaire, A. J., Garcia-Romeu, M. L., & Vallellano, C. (2014). Critical analysis of necking and fracture limit strains and forming forces in single-point incremental forming. *Materials and Design, 63*, 20–29.

Centeno, G., Silva, M. B., Cristino, V. A. M., Vallellano, C., & Martins, P. A. F. (2012). Hole-flanging by incremental sheet forming. *International Journal of Machine Tools and Manufacture, 59*, 46–54.

Eberle, B., Volk, W., & Hora, P. (2008). Automatic approach in the evaluation of the experimental FLC with a full 2D approach based on a time depending method. In *Presented at the 7th Numisheet conference and workshop, Interlaken, Switzerland* (pp. 279–284).

Embury, J. D., & Duncan, J. L. (1981). Formability maps. *Annual Review of Materials Science, 11*, 505–521.

Feldmann, P., Schatz, M., & Aswendt, P. (2009). Automatic FLC-value determination from 4D strain data. In *Presented at the International Deep Drawing Research Group IDDRG, Golden, USA* (pp. 533–546).

Geiger, M., & Merklein, M. (2003). Determination of forming limits diagrams—A new analysis method for characterization of materials formability. *CIRP Annals – Manufacturing Technology, 52*(1), 213–216.

Gurson, A. (1977). Continuum theory of ductile rupture by void nucleation and growth. I. Yield criteria and flow rules for porous ductile media. *Journal of Engineering Materials and Technology, Transactions of the ASME, 99*, 2–15.

Hill, R. (1948). A theory of yielding and plastic flow of anisotropic metals. *Proceedings of the Royal Society of London, Series A, 193*, 281–297.

Hill, R. (1952). On discontinuous plastic states with special reference to localized necking in thin sheets. *Journal of the Mechanics and Physics of Solids, 1*, 19–30.

Hotz, W., Merklein, M., Kuppert, A., Friebe, H., & Klein, M. (2013). Time dependent FLC determination—Comparison of different algorithms to detect the onset of unstable necking before fracture. *Key Engineering Materials, 549*, 397–404.

Hotz, W., & Timm, J. (2008). Experimental determination of forming limit curves (FLC). In *Presented at the 7th Numisheet conference and workshop, Interlaken, Switzerland* (pp. 271–278).

Huang, M., Zhang, L., & Yang, L. (2008). On the failure of AHSS at tooling radius. In *Presented at the 7th Numisheet conference and workshop (workshop on numerical simulation of 3D sheet metal forming processes), Interlaken, Switzerland* (pp. 307–309).

International Standard ISO 12004-2:2008 *Metallic materials-sheet and strip-determination of forming limit curves, Part 2: Determination of forming limit curves in the laboratory.* (2008). Geneva, Switzerland: International Organisation for Standardization.

Isik, K., Silva, M. B., Tekkaya, A. E., & Martins, P. A. F. (2014). Formability limits by fracture in sheet metal forming. *Journal of Materials Processing Technology, 214*, 1557–1565.

Li, J., Carsley, J. E., Stoughton, T. B., Hector, L. G., & Hu, S. J. (2013). Forming limit analysis for two-stage forming of 5182-O aluminum sheet with intermediate annealing. *International Journal of Plasticity, 45*, 21–43.

Li, Y., Luo, M., Gerlach, J., & Wierzbicki, T. (2010). Prediction of shear-induced fracture in sheet metal forming. *Journal of Materials Processing Technology, 210*, 1858–1869.

Marciniak, Z. (1965). Stability of plastic shells under tension with kinematic boundary condition. *Archiwum Mechaniki Stosorwanej, 17*, 577–592.

Martínez-Donaire, A. J., García-Lomas, F. J., & Vallellano, C. (2014). New approaches to detect the onset of localized necking in sheets under through-thickness strain gradients. *Materials & Design, 57*, 135–145.

Martínez-Donaire, A. J., Vallellano, C., Morales, D., & García-Lomas, F. J. (2009). In *On the experimental detection of necking in stretch-bending tests: Vol. CP1181* (pp. 500–508). College Park, MD: American Institute of Physics.

Martínez-Donaire, A. J., Vallellano, C., Morales, D., & García-Lomas, F. J. (2010). Experimental detection of necking in stretch-bending conditions: A critical review and new methodology. *Steel Research International, 81*(9), 785–788.

Martins, P. A. F., Bay, N., Tekkaya, A. E., & Atkins, A. G. (2014). Characterization of fracture loci in metal forming. *International Journal of Mechanical Sciences, 83*, 112–123.

McClintock, F. A. (1968). A criterion for ductile fracture by the growth of holes. *Journal of Applied Mechanics, Transactions of the ASME, 35*, 363–371.

Merklein, M., Kuppert, A., & Geiger, M. (2010). Time dependent determination of forming limit diagrams. *CIRP Annals – Manufacturing Technology, 59*, 295–298.

Morales, D., Martínez-Donaire, A. J., Vallellano, C., & García-Lomas, F. J. (2009). Bending effect in the failure of stretch-bend metal sheets. *International Journal of Material Forming, 2*, 813–816.

Morales-Palma, D., Vallellano, C., & García-Lomas, F. J. (2013). Assessment of the effect of the through-thickness strain/stress gradient on the formability of stretch-bend metal sheets. *Materials & Design, 50*, 798–809.

Muscat-Fenech, C. M., Arndt, S., & Atkins, A. G. (1996). The determination of fracture forming limit diagrams from fracture toughness. In *Presented at the 4th international conference. Sheet metal 1996 Vol. 1* (pp. 249–260). The Netherlands: University of Twente.

Nahshon, K., & Hutchinson, J. (2008). Modification of the Gurson model for shear failure. *European Journal of Mechanics A/Solids, 27*, 1–17.

Sène, N. A., Balland, P., Arrieux, R., & Bouabdalah, K. (2012). An experimental study of the microformability of very thin materials. *Experimental Mechanics, 53*(2), 155–162.

Silva, M. B., Nielsen, P. S., Bay, N., & Martins, P. A. F. (2011). Failure mechanisms in single point incremental forming of metals. *International Journal of Advanced Manufacturing Technology, 56*, 893–903.

Silva, M. B., Skjoedt, M., Atkins, A. G., Bay, N., & Martins, P. A. F. (2008). Single point incremental forming & formability/failure diagrams. *Journal of Strain Analysis for Engineering Design, 43*, 15–36.

Silva, M. B., Skjoedt, M., Bay, N., & Martins, P. A. F. (2009). Revisiting single-point incremental forming and formability/failure diagrams by means of finite elements and experimentation. *Journal of Strain Analysis for Engineering Design, 44*(4), 221–234.

Situ, Q., Jain, M., & Bruhis, M. A. (2006). Suitable criterion for precise determination of incipient necking in sheet materials. *Material Science Forum, 519–521*, 111–116.

Situ, Q., Jain, M. K., & Bruhis, M. (2007). Further experimental verification of a proposed localized necking criterion. In *Presented at the numerical methods in industrial forming processes—9th Numiform, Porto, Portugal* (pp. 907–912).

Swift, H. W. (1952). Plastic instability under plane stress. *Journal of the Mechanics and Physics of Solids, 1*, 1–18.

Tharret, M. R., & Stoughton, T. B. (2003). *Stretch-bend forming limits of 1008 AK steel.* Technical paper no. 2003-01-1157 SAE International.

Tvergaard, V., & Needleman, A. (1984). Analysis of the cup-cone fracture in a round tensile bar. *Acta Metallurgica, 32*, 157–169.

Volk, W., & Hora, P. (2010). New algorithm for a robust user-independent evaluation of beginning instability for the experimental FLC determination. *International Journal of Material Forming, 4*, 339–346.

Wang, K., Carsley, J. E., He, B., Li, J., & Zhang, L. (2014). Measuring forming limit strains with digital image correlation analysis. *Journal of Materials Processing Technology, 214*, 1120–1130.

Hole-flanging by single point incremental forming

2

M.B. Silva[1], N. Bay[2], P.A.F. Martins[1]
[1]Universidade de Lisboa, Lisboa, Portugal; [2]Technical University of Denmark, Lyngby, Denmark

2.1 Introduction

Hole-flanging is a conventional metal forming process in which a sheet blank (hereafter named as "blank"), with the outer periphery rigidly fixed by a blank holder, is forced by a punch and die to produce a cylindrical, rectangular, or complex-shaped flange in a blank with a precut hole (Figure 2.1a).

The process is carried out in press-working operations running in either compounded or progressive tools and is widely used to strengthen the edge of the holes, to improve its appearance, or to provide additional support for joining sheets to tubes and profiles, among other applications.

Hole-flanging produced by conventional press-working involves local plastic deformation by bending and stretching and has been extensively investigated since the late 1960s. Pioneering studies by Yamada and Koide (1968) applied the incremental theory of plasticity to analyze the combined effect of the yield stress and strain hardening in the distribution of strains and stresses. Subsequent work by Wang and Wenner (1974) revealed that the state of stress in the flange is dominantly uniaxial and established a closed-form expression, based on the total strain membrane theory, to calculate the maximum allowable strain as a function of the geometry of the hole and flange. Johnson, Chitkara, and Minh (1977) investigated the fundamentals of plastic flow leading to failure in hole-flanging and concluded that the onset of fracture is strongly influenced by the geometry and the mechanical properties of the material, namely, the plastic anisotropy ratio.

A literature review of sheet metal forming performed by Mackerle (2004) allows the conclusion that the process window of hole-flanging produced by conventional press-working is characterized by the limiting forming ratio (LFR), above which material failure can take place. The LFR is defined as the ratio of the maximum inside diameter D_{max} of the finished flange to the initial diameter D_0 of the precut hole (Figure 2.1c),

$$\text{LFR} = \frac{D_{max}}{D_0} \tag{2.1}$$

and its value is dependent on the mechanical properties of the material, surface quality of the hole edge, geometry of the forming punch, clearance between the punch and die, and lubrication conditions (Stachowicz, 2008).

The literature review also allows concluding that hole-flanging produced by conventional press-working requires large batch sizes to justify the investment in tools

Materials Forming and Machining. http://dx.doi.org/10.1016/B978-0-85709-483-4.00002-8

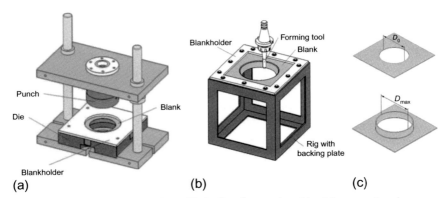

Figure 2.1 Schematic representation of hole-flanging produced by (a) conventional press-working and (b) single point incremental forming (SPIF) together with a drawing of (c) a cylindrical flange produced in a blank with a circular precut hole.

and other major equipment, which are necessary to ensure forming in a single press stroke in compounded tools or in several strokes in progressive tools.

Current changes and prospects for the metalworking industry requiring very small life-cycles with short conception and design lead times has been encouraging the development of new manufacturing processes that are capable of reducing the costs to a level where small-batch sizes become economically feasible.

Rapid tools made from low-cost polymers and composites or special sheet metal forming processes like spinning or hydromechanical forming are utilized in rapid prototyping and small-batch production of sheet metal parts, but they all suffer from technical or economic constraints. Rapid tools have much lower strength than conventional punches and dies, and their use is often limited by premature fatigue failure or by poor tolerances of the resulting parts. In addition, the mechanisms by which rapid tools fail are also not fully understood (Park & Colton, 2003). Spinning and related processes are limited to axisymmetric parts and may involve significant labor costs unless operations are automated (Wong, Dean, & Lin, 2003). Hydromechanical forming allows producing complex and high-quality sheet metal parts but its utilization is limited by the overall high equipment costs (Lang et al., 2004).

Still, all the above-mentioned processes are commonly used, for example, to produce special and small batches of sheet metal parts for the aerospace industry and stainless steel parts for home appliances and the food industry.

Focusing the goal on the production of flanges in sheet metal parts and taking into account the actual needs for rapid prototyping and flexible manufacturing applied to the widest possible range of materials, shapes, and applications, it seems that single point incremental forming (SPIF) is one of the emerging fabrication processes that is capable of reducing the costs to a level where small-batch production of hole-flanged parts becomes economically feasible.

The state-of-the-art and potential applications of SPIF are comprehensively described in the state-of-the-art review papers by Jeswiet et al. (2005) and Nimbalkar and Nandedkar (2013). Hole-flanging produced by SPIF is a recent variant of the process in which a blank with a precut hole is progressively shaped into a hole-flanged part with a forming tool provided with a hemispherical tip mounted in a numerically controlled (CNC) machining center and following a path generated by computer-aided manufacturing (CAM) software. The basic components of hole-flanging produced by SPIF are schematically illustrated in Figure 2.1b and comprise: (i) the blank with a precut hole, (ii) the rig with the backing plate (not seen in the figure), (iii) the blank holder, and (iv) the hemispherical-ended forming tool (hereafter referred to as "the tool pin"). The periphery of the blank is placed on top of the backing plate and is rigidly clamped around its edges by the blank holder. The remaining surface of the blank is unsupported underneath and is progressively shaped by the free or forced rotating tool pin describing the contour of the final desired geometry.

The first research work in hole-flanging produced by SPIF was performed by Cui and Gao (2010) who studied the influence of different multistage forward tool path trajectories on the formability limits of AA1060 aluminum blanks. Subsequent work by Petek, Kuzman, and Fijavž (2011) was focused on the feasibility of employing multistage backward tool path trajectories to produce symmetrical and asymmetrical flanges in DC05 mild steel blanks with precut holes. Centeno, Silva, Cristino, Vallellano, and Martins (2012) and Silva, Teixeira, Reis, and Martins (2013) studied the deformation mechanics of the process and the physics behind the occurrence of failure in hole-flanging of aluminum AA1050 and titanium grade 2 blanks. In their paper, Silva et al. (2013) also showed that the LFR of hole-flanging produced by SPIF is only more favorable than that of hole-flanging produced by conventional press-working when the fracture forming limit line (FFL) is placed well above the forming limit curve (FLC). Montanari, Cristino, Silva, and Martins (2013) revisited the deformation mechanics of the process and presented the loading paths in the principal strain and stress spaces.

Despite the growing research interest on hole-flanging produced by SPIF, the large majority of the existing publications are primarily and almost exclusively focused on conical and cylindrical flanges obtained from blanks with precut circular holes. The only exceptions, as far as the authors are aware, are the publications authored by Petek et al. (2011), Voswinckel, Bambach, and Hirt (2013), and Cristino, Montanari, Silva, and Martins (2014).

Petek et al. (2011) showed that significant trial and error procedures are needed to successfully produce asymmetrical (non-round) flanges by SPIF and claimed that this is due to inappropriate definition of the initial shape of the precut hole and to different deformation histories along the walls. They concluded that research in hole-flanging produced by SPIF should be directed toward the challenge of producing asymmetrical hole-flanges with minimal trial and error procedures.

Voswinckel et al. (2013) investigated the feasibility of fabricating stretch and shrink flanges at the sheet edges by multistage SPIF. They studied the influence of tool path strategies and concluded that the feasibility ratio of flange length to flange radii could distinctly exceed those of conventional flanging produced by press-working.

Cristino et al. (2014) investigated the deformation mechanics of square flanges with round corners produced by SPIF and concluded that the strain envelopes of the grid points located at the transition between the sides and the corners are similar to those found in cylindrical hole-flanges with large radii whereas the strain envelopes of

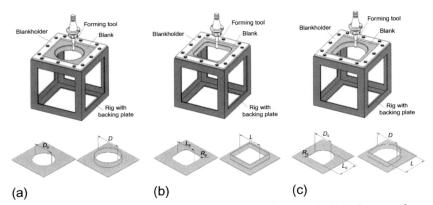

Figure 2.2 Schematic representation of hole-flanging produced by SPIF in the case of
(a) cylindrical hole-flanged parts, (b) square hole-flanged parts with round corners, and
(c) complex hole-flanged parts combining features of cylindrical and square flanges.

the grid points located in between the middle of the corners and the transition between
the sides and the corners are completely different due to nonlinear strain loading conditions that evolve from near plane strain into uniaxial tension.

From what was mentioned before, it can be concluded that there is a need to revisit
and systematize the fundamentals of hole-flanging produced by SPIF in a publication
that draws from existing results on cylindrical and square flanges to new results on
complex flanges (Figure 2.2).

This chapter is intended to meet this objective by focusing on the deformation mechanics and on the physics behind the occurrence of failure in hole-flanging produced
by SPIF. In specific terms, the chapter is aimed to give answers to the following main
questions: What is the influence of the precut hole geometry on the overall formability of the process? Which are the main differences in plastic flow arising from the
interaction between the hemispherical tool tip and the blank in different locations of
the parts? Which is the physics behind the occurrence of failure in axisymmetric and
asymmetric hole-flanging produced by SPIF? What is the feasibility of setting-up process formability windows as a function of the main operating parameters?

The answers to these questions are given by means of research methodoloies that
combine circle-grid analysis, anisotropic plasticity, mechanical characterization of
the material, and fabrication of flanges by multistage SPIF in blanks with different
precut holes.

2.2 Materials and methods

2.2.1 Mechanical characterization

The work was carried out on aluminum AA1050-H111 sheets with 1 mm thickness,
and the mechanical characterization of the material at room temperature was performed by means of tensile tests on an INSTRON 4507 universal testing machine.

The tests followed the ASTM Standard E8/E8M – 09 (2013) and the resulting average stress–strain curve was approximated by the following Ludwik–Hollomon's equation:

$$\sigma = 140\varepsilon^{0.04} \ (\text{MPa})$$

(2.2)

The normal $\bar{r}$ and planar Δr anisotropy coefficients were determined from,

$$\bar{r} = \frac{r_0 + 2r_{45} + r_{90}}{4}, \quad \Delta r = \frac{r_0 - 2r_{45} + r_{90}}{2}$$

(2.3)

where r_0, r_{45}, and r_{90} are the anisotropy coefficients obtained from specimens cut out from the supplied sheets at 0°, 45°, and 90° with respect to the rolling direction. Table 2.1 provides a summary of the mechanical properties of aluminum AA1050-H111.

2.2.2 Formability characterization

The formability limits at necking (FLC) and fracture (fracture forming limit line [FFL]) were characterized by means of laboratory tests that covered strain paths from uniaxial to plane-strain and biaxial loading conditions. The sheet formability tests that were used by the authors are schematically listed in Table 2.2.

The Nakazima and hemispherical dome tests were performed in a flexible tool system that was installed in the INSTRON 4507 universal testing machine where the tensile tests were performed, whereas the circular and elliptical hydraulic bulge tests were performed in an ERICHSEN 145/60 hydraulic universal testing machine. The specimens utilized in all the aforementioned sheet formability tests were electrochemically etched with a grid of overlapping circles with a 2 mm initial diameter d.

The methodology used for determining the FLC was based upon measuring the in-plane strains $(\varepsilon_1, \varepsilon_2)$ from grid points located along predefined directions crossing the crack perpendicularly by means of a computer-aided measuring system (Figure 2.3a). The in-plane strains $(\varepsilon_1, \varepsilon_2)$ at the grid points were obtained as follows:

$$\varepsilon_1 = \ln\left(\frac{a}{d}\right), \quad \varepsilon_2 = \ln\left(\frac{b}{d}\right)$$

(2.4)

Table 2.1 Summary of the mechanical properties of aluminum AA1050-H111 sheets with 1 mm thickness

	Modulus of elasticity (GPa)	Yield strength (MPa)	Ultimate tensile strength (MPa)	Elongation at break (%)	Anisotropy coefficient
0° RD	72.7	115.4	119.0	7.1	0.71
45° RD	67.9	120.4	121.2	5.2	0.88
90° RD	71.8	123.0	120.8	5.6	0.87
Average	70.0	119.9	120.5	6.8	$\bar{r} = 0.84$
					$\Delta r = -0.11$

Table 2.2 **Sheet formability tests that were utilized for determining the FLC and FFL of aluminum AA1050-H111 sheets with 1 mm thickness**

Test	Deformation mode	State of strain	State of stress	Schematic drawing
Tensile test	Uniaxial	$\varepsilon_1 > 0$ $\varepsilon_2 = \varepsilon_3 < 0$ $\varepsilon_2 = \varepsilon_3 = -\varepsilon_1 / 2$	$\sigma_1 > 0$ $\sigma_2 = \sigma_3 = 0$	
Hecker's variant of Nakazima test	Between uniaxial and biaxial	$\|\varepsilon_1 > 0\|$ $-\varepsilon_1 / 2 < \varepsilon_2 < \varepsilon_1$ $\varepsilon_3 = -(\varepsilon_1 + \varepsilon_2)$	$\sigma_1 > 0$ $\sigma_1 > \sigma_2 > 0$ $\sigma_3 = 0$	
Hemi spherical dome test	Biaxial	$\varepsilon_1 = \varepsilon_2 > 0$ $\varepsilon_3 < 0$ $\varepsilon_1 = \varepsilon_2 = -\varepsilon_3 / 2$	$\sigma_1 = \sigma_2 > 0$ $\sigma_3 = 0$	
Hydraulic bulge test				

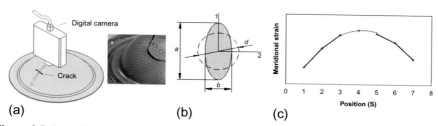

Figure 2.3 Experimental procedure for determining the strain pairs at the onset of necking from measurements in adjacent deformed circles along a direction perpendicular to the crack. (a) Schematic drawing and picture of a test specimen, (b) typical ellipse of a grid point, and (c) the "bell-shaped curve" of the interpolation procedure.

where a and b are the lengths of the major and minor axes of the ellipses that resulted from plastic deformation of the original grid of overlapping circles during sheet formability tests (Figure 2.3b).

The FLC was obtained by interpolation of the experimental in-plane strains (2.4) into a "bell-shaped curve" in order to reconstruct the distribution of strains in the area of intense localization and by subsequent extrapolation of the maximum strain pairs at the onset of necking. The procedure is schematically shown in Figure 2.3c and the strain pairs at the onset of necking for the aluminum AA1050-H111 are given by the gray open markers in Figure 2.4.

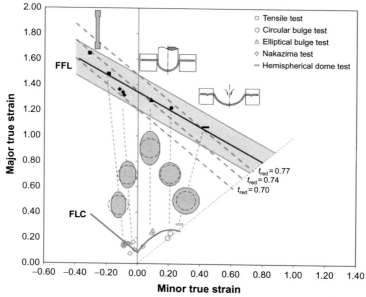

Figure 2.4 Forming limit curve (FLC) and fracture forming limit line (FFL) of aluminum AA1050-H111 in the principal strain space. The black solid markers correspond to failure by fracture, the gray open markers are associated to the onset of necking and the gray vertical dashed lines represent the change in strain path direction toward plane strain after necking.

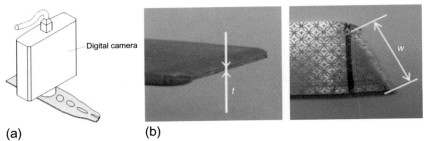

(a)   (b)

Figure 2.5 (a) Conventional procedure for determining the strain pairs from measurements in adjacent deformed circles along a direction perpendicular to the crack and (b) recommended procedure for determining the strain pairs at the onset of fracture from measurements of strain along the thickness and the width directions with corresponding pictures of a tensile test specimen.

In contrast, the methodology used for determining the FFL was based upon measuring the thickness of the specimens before and after fracture at several locations along the crack in order to obtain the "gauge length" strains. This is because the application of grids, even with very small circles in order to obtain strains in the necking region after it forms and, therefore, close to the fracture, provides strain values that cannot be considered the fracture strains (Figure 2.5a). Moreover, such grids create measurement problems and suffer from sensitivity to the initial size of the circles used in the grids due to the inhomogeneous deformation in the neighborhood of the crack.

The procedure is schematically shown in Figure 2.5b and the resulting strain pairs at the onset of fracture for aluminum AA1050-H111 are given by the black solid markers that were interpolated by means of the straight line (FFL) falling from left to right in Figure 2.4,

$$\varepsilon_1 + 0.79\varepsilon_2 = 1.37 \tag{2.5}$$

This line is in fair agreement with the theoretical condition of critical thickness reduction at fracture (corresponding to a straight line with slope "−1") that is comprehensively discussed in references (Atkins, 1996; Isik, Silva, Tekkaya, & Martins, 2014). The gray area around the FFL in Figure 2.4 corresponds to an uncertainty interval of 10% associated with its experimental determination.

The iso-thickness reduction lines $t_{red} = (t_0 - t)/t_0$, where t_0 and t are the original (undeformed) and the actual sheet thicknesses that are plotted in Figure 2.4 were determined from the condition of constant thickness reduction at fracture (Atkins, 1996),

$$t_{red} = 1 + e^{(\varepsilon_1 + \varepsilon_2)} \tag{2.6}$$

As seen in Figure 2.4, the reduction in sheet thickness at the onset of fracture corresponds to maximum values $t_{red} > 0.7$. Further details on the methods and procedures used to characterize the FLC and the FFL can be found elsewhere (Silva, Nielsen, Bay, & Martins, 2011).

2.2.3 Hole-flanging experiments

The experiments in hole-flanging produced by SPIF were carried out in a Deckel Maho CNC machining center equipped with a rig similar to that schematically shown in Figure 2.2. The work plan was designed to provide a wide range of experimental data and observations for investigating the influence of the flange geometry, size and shape of the precut holes, and diameter of the hemispherical tool tips on plastic flow and failure of the hole-flanged parts produced by multistage SPIF.

The flanges were fabricated from aluminum AA1050-H111 blanks 250 mm × 250 mm × 1 mm that were milled to deliver precut holes of different sizes and shapes. The edges of the precut holes were subsequently ground with grit sand paper to eliminate burrs and cracks. After finishing preparing the precut holes, the blanks were electrochemically etched with grids of circles with a 2.5 mm initial diameter in order to allow in-plane strains to be measured from the deformed ellipses by means of circle-grid analysis.

The experiments were divided into four different groups that are summarized in Tables 2.3–2.6. Table 2.3 refers to cylindrical flanges with varying precut holes that were formed by means of a multistage forward tool path strategy and using progressively increasing drawing angles from $\psi_1 = 65°$, until $\psi_6 = 90°$ with steps of $\Delta\psi = 5°$ in order to obtain vertical walls. The test cases listed in this table were utilized for investigating plastic flow and failure by means of the extended circle-grid analysis to be presented in Section 2.3 that allows tracing the strain and stress loading paths of the grid points during hole-flanging produced by multistage SPIF.

Table 2.4 includes the set of experiments that were performed with the purpose of studying the influence of the tool tip diameters on the process window of cylindrical hole-flanged parts.

Table 2.5 refers to square flanges with round corners that were formed by means of a multistage forward tool path strategy identical to that used for cylindrical flanges. The test cases included in this table were aimed at investigating the influence of the precut hole geometry on plastic flow and failure of square flanges with round corners. The tool diameter was kept constant, because its influence on the overall formability limits of hole-flanged parts had been already addressed in case of cylindrical flanges (Table 2.4). This allowed the investigation on square flanges with round corners to be focused on the design of the precut holes and on the deformation mechanics of the different regions of the inclined walls.

Finally, Table 2.6 refers to the set of experiments that were performed in order to investigate the formability of complex flanges with combined features of cylindrical and square flanges.

The tool pins used in the experiments were made in cold working 120WV4-DIN tool steel, hardened and tempered to 60 HRc. The tool paths were generated with the commercial software MasterCAM using the options available for spherical ball nose end mills (e.g., software compensation) and consisted of helical trajectories with a feed rate of 1000 mm/min and a downward feed size of 0.2 mm per revolution. The main reason for keeping the step size and feed rate constant in all tests was that these parameters had already been investigated in conventional SPIF (Hirt, Ames, Bambach, & Kopp, 2004; Silva et al., 2011) and cylindrical hole-flanging produced by SPIF (Centeno et al., 2012; Cui & Gao, 2010).

Table 2.3 Work plan for cylindrical flanges produced by SPIF

Precut hole diameter, D_0 (mm)	Tool diameter (mm)	Backing plate diameter (mm)	Drawing angle of the intermediate stages, ψ_i (°)						
			65	70	75	80	85	90	
102 111 114 120	8.0	148.5							

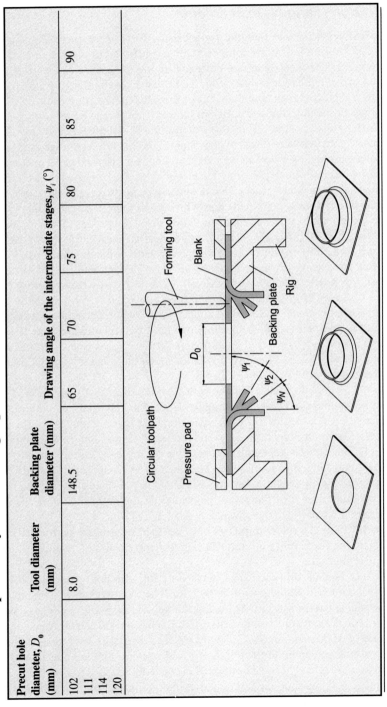

Table 2.4 **Work plan for cylindrical flanges produced by SPIF with varying tool tip diameters**

Precut hole diameter, D_0 (mm)	Tool diameter (mm)	Backing plate diameter (mm)	
100 110 120 130	8, 12, 25	165.0	

Table 2.5 **Work plan for square flanges with round corners produced by SPIF**

Precut hole geometry		Tool diameter (mm)	Backing plate side length (mm)	
Side length, L_0 (mm)	Corner radius, R_0 (mm)			
126 136 140 147 154	5, 10, 22.5, 31.5	8	170	

Table 2.6 **Work plan for complex flanges produced by SPIF**

Precut hole geometry		Tool diameter (mm)	Backing plate diameter (mm)	Backing plate side length (mm)	
Hole diameter, D_0 (mm)	Corner radius, R_0 (mm)				
128	10, 15, 20, 30	8	170	170	

The rotation of the tool was free because previous experiments by Silva, Skjoedt, Bay, and Martins (2009) showed that the overall formability of SPIF is not significantly influenced by the rotation of the forming tool due to the fact that circumferential friction resulting from the contact between the tool and the blank is negligible.

The lubricant utilized in the experiments was the forming fluid Castrol Iloform TDN81.

2.3 Extended circle-grid analysis

Extended circle-grid analysis is a new approach for tracing the deformation history of material elements in sheet metal forming processes carried out under proportional strain paths with a slope defined by β (Montanari et al., 2013),

$$\beta = \frac{d\varepsilon_2}{d\varepsilon_1} = \frac{\varepsilon_2}{\varepsilon_1} \tag{2.7}$$

The approach is built upon a combination of circle-grid analysis used for measuring the in-plane strains $(\varepsilon_1, \varepsilon_2)$ from the major and minor axis of the ellipses that result from the plastic deformation of the grid points with the generalization to anisotropic plastic flow of the isotropic framework conditions assumed by Glover, Duncan, and Embury (1977) in their overall investigation on failure maps in sheet metal forming.

The application of extended circle-grid analysis to hole-flanging produced by SPIF allows determining the strains, stresses, and ductile damage along the intermediate stages of deformation and comparing their maximum achievable values against the onsets of necking and fracture in the principal strain and stress spaces. The assumption of proportional strain paths allows application of extended circle-grid analysis to hole-flanging of cylindrical flanges as will later be proved experimentally.

Assuming that the average value of the normal anisotropy coefficient $\bar{r}$ is constant during straining and considering the anisotropic yield criterion by Hill (1948) under plane stress conditions $\sigma_3 = 0$, the associated flow rule relating the in-plane strain increments with the applied stresses is expressed as (in what follows, $\bar{r}$ will be simply written as r),

$$d\varepsilon_1 = \frac{d\bar{\varepsilon}}{\bar{\sigma}}\left[\frac{1}{1+r}\right]\left(\sigma_1 + r\left(\sigma_1 - \sigma_2\right)\right), \quad d\varepsilon_2 = \frac{d\bar{\varepsilon}}{\bar{\sigma}}\left[\frac{1}{1+r}\right]\left(\sigma_2 + r\left(\sigma_2 - \sigma_1\right)\right) \tag{2.8}$$

where the effective strain increment $d\bar{\varepsilon}$ and the effective stress $\bar{\sigma}$, respectively, are calculated from

$$d\bar{\varepsilon} = \frac{1+r}{\sqrt{(1+2r)}}\sqrt{d\varepsilon_1^2 + d\varepsilon_2^2 + \frac{2r}{(1+r)}d\varepsilon_1 d\varepsilon_2} \tag{2.9}$$

and

$$\bar{\sigma} = \sqrt{\sigma_1^2 + \sigma_2^2 - \frac{2r}{(1+r)}\sigma_1\sigma_2} \tag{2.10}$$

Now, determining the ratio $\beta = d\varepsilon_2 / d\varepsilon_1$ between the minor and major strain increments (2.7) and assuming proportional loading conditions with a slope $\alpha = \sigma_2 / \sigma_1$, one obtains

$$\alpha = \frac{\sigma_2}{\sigma_1} = \frac{(1+r)\beta + r}{(1+r) + r\beta} \tag{2.11}$$

By inserting (2.11) into (2.10), the following expressions for the principal in-plane stresses σ_1 and σ_2 are obtained,

$$\sigma_1 = \frac{\bar{\sigma}}{\sqrt{\left(1 - \frac{2r}{1+r}\alpha + \alpha^2\right)}}, \quad \sigma_2 = \frac{\alpha\bar{\sigma}}{\sqrt{\left(1 - \frac{2r}{1+r}\alpha + \alpha^2\right)}} \tag{2.12}$$

where $\bar{\sigma}$ is calculated from the stress–strain curve of the material (2.2) after integrating the increment of effective strain $d\bar{\varepsilon}$ up to the current plastic deformation $\bar{\varepsilon} = \int d\bar{\varepsilon}$.

Equation (2.12) allows determining the major and minor in-plane stresses that are needed to plot the resulting stress loading paths in the principal stress space, directly from the experimental strain pairs $(\varepsilon_1, \varepsilon_2)$ at the grid points. The application of the overall procedure will later be shown in case of cylindrical hole-flanges.

The evolution of accumulated damage D and the determination of the critical value of accumulated damage D^{crit} at the location of crack initiation from the beginning until the onset of failure are easily resolved from the strain and stress values that were previously obtained from the above described extended circle-grid analysis. In case of hole-flanging produced by SPIF, the damage function used by the authors is based on the stress triaxiality ratio $\sigma_m / \bar{\sigma}$ that was originally proposed by McClintock (Martins, Bay, Tekkaya, & Atkins, 2014),

$$D = \int_0^{\bar{\varepsilon}} \frac{\sigma_m}{\bar{\sigma}} d\bar{\varepsilon} \tag{2.13}$$

where σ_m is the average stress and $\bar{\varepsilon}$ is the effective strain at a specific location.

The choice of the damage function (2.13) was made by combining the work of Muscat-Fenech, Arndt, and Atkins (1996), who were able to correlate the FFL with fracture toughness in mode I, with the experimental observations of Silva, Skjoedt, Atkins, Bay, and Martins (2008), who concluded that failure by fracture in SPIF occurs by crack opening in mode I due to meridional stresses that are applied along the plastically deforming region resulting from the contact between the sheet and the hemispherical tool tip.

In case of a location corresponding to crack initiation, the effective strain $\bar{\varepsilon} = \bar{\varepsilon}_f$ and the accumulated values of damage given by (2.13) becomes the critical value of damage D^{crit} at the onset of failure. The application to hole-flanging of cylindrical flanges produced by SPIF will be discussed in the following section.

2.4 Results and discussion

2.4.1 Cylindrical flanges

2.4.1.1 Plastic flow and failure

The deformation mechanics of cylindrical flanges produced by multistage SPIF was characterized by tracing the strain and stress loading paths of selected grid points (A–J) located along the meridional direction of the blanks with circular precut holes at different intermediate stages of deformation (refer to the intermediate drawing angles in Table 2.3 and to Figure 2.6).

Considering the blanks with a precut hole diameter $D_0 = 120$ mm, the strain paths of points A–J at the six intermediate drawing angles ($\psi_i = 65$–$90°$) grow linearly and monotonically from the origin to the maximum achievable strains (refer to the thin black lines in Figure 2.6a). There are no changes in the strain paths at the transition from the FLC toward the FFL in contrast to what was previously observed in the sheet formability tests that were presented in Table 2.2 (refer to the gray vertical dashed lines in Figure 2.4).

The thick black curve in Figure 2.6a is the strain envelope of the greatest achievable strains for any linear strain path in hole-flanging produced by SPIF. The points (I, J) located near the hole edge undergo equal biaxial stretching, the points (F–H) located at the transition region between the flange wall and the hole edge evolve under biaxial stretching, and the points located at the corner (A, B) and at the flange wall (C–E) experience near plane-strain conditions.

The shape of the envelope clearly indicates the critical region where strains are larger and closer to the FFL to be located at the flange wall (refer to points C–E) and not at the hole edge as it is commonly found in hole-flanging by conventional press-working. This is because the material placed at the vicinity of the hole edge deforms under biaxial stretching ($d\varepsilon_\varphi, d\varepsilon_\theta > 0$) instead of uniaxial tension (as in the case of hole-flanging by conventional press-working) due to a combination of plane-strain deformation ($d\varepsilon_\varphi : d\varepsilon_\theta : d\varepsilon_t = 1 : 0 : -1$) typical of incremental sheet metal forming and uniaxial tension ($d\varepsilon_\varphi : d\varepsilon_\theta : d\varepsilon_t = -1/2 : 1 : -1/2$) associated with the progressive increase of the inner diameter of the hole.

The stress paths and the stress envelope corresponding to the greatest achievable stresses and the locus of the stress states corresponding to the FLC and the FFL are plotted in Figure 2.6b and were determined by means of Equation (2.12).

The analysis of Figure 2.6a and b allows concluding that point C is the closest to failure but still allowing the fabrication of a sound part from a hole-flanged blank with a precut hole diameter $D_0 = 121$ mm (refer to the picture in Figure 2.6a).

In contrast, the strain envelope corresponding to the second intermediate drawing angle $\psi_i = 70°$ in blanks with a precut hole diameter $D_0 = 102$ mm (Table 2.3) presents a maximum within the region of failure by fracture in good agreement with the experimental observations, which indicates that cracking occurs in opening mode I due to meridional tensile stresses (Figure 2.7).

The contours of accumulated damage obtained from Equation (2.13) allow determining a critical damage $D^{\text{crit}} = 0.81$ at the onset of failure by cracking (Figure 2.7).

2.4.1.2 Process window

Table 2.7 offers a complete overview of the results obtained in the production of cylindrical flanges by multistage SPIF using blanks with different precut hole diameters D_0 and various tool tip diameters. As seen from the table, it is not possible to complete the hole-flanging process without failure (i.e., it is not possible to shape the desired flanges with vertical walls) in blanks with precut hole diameters $D_0 = 102$ mm due to the occurrence of circumferential cracking.

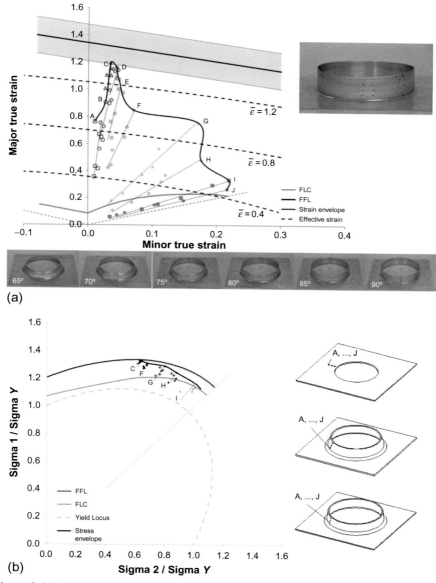

Figure 2.6 (a) Strain loading paths in the principal strain space and (b) stress loading paths in the principal stress space of the grid points A–J located along the meridional direction of a cylindrical hole-flanged part with a precut hole diameter $D_0 = 120$ mm at drawing angles $\psi_i = 65$–$90°$. *Note*: The dashed black elliptical curves in the principal strain space are the iso-effective strain contours and the dashed gray elliptical curve in the principal stress space is the yield locus.

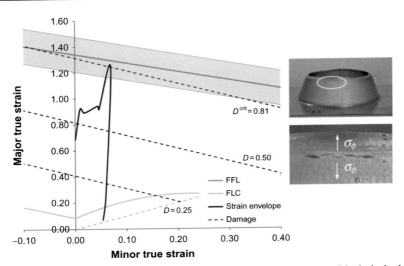

Figure 2.7 Strain envelope in the principal strain space with line contours (black dashed lines) showing the iso-values of accumulated damage D and the critical damage D^{crit} in case of a cylindrical flanged part made from a blank with a precut hole diameter $D_0 = 102\,\text{mm}$ at the onset of failure by cracking.

On the other hand, the minimum precut hole diameter is also found to increase as the tool diameter increases. In fact, when the tool diameter increases from 8 to 12 mm it is no longer possible to successfully shape a blank with a precut hole diameter $D_0 = 120$ mm into a cylindrical hole-flanged part, because the maximum admissible angle is found to decrease from $\psi_6 = 90°$ to $\psi_2 = 85°$. Instead, a blank with a precut hole diameter $D_0 = 130$ mm needs to be used. The reduction of the maximum admissible drawing angle ψ_i with the increase of the tool diameter is in close agreement with the reduction of formability with increasing tool diameter that was previously observed in conventional SPIF by Silva et al. (2009).

The results for cylindrical flanges produced by SPIF with a tool diameter of 25 mm may seem in disagreement with what was mentioned before but the "apparent" increase in formability results from the fact that the deformation state changes from plane strain into tensile as the tool diameter increases from 12 to 25 mm due to forming conditions more similar to those found in hole-flanging produced by conventional press-working. This is shown in Figure 2.8 where it is also possible to observe a reduction in the total height of the flange with the increase of the tool diameter.

2.4.2 Square flanges with round corners

2.4.2.1 Plastic flow and failure

The deformation mechanics of square flanges with round corners is investigated by breaking down the final geometry of the flange into four corners, each of them resembling a quarter of a cylindrical flange with a very small radius (refer to the light gray

Table 2.7 Summary of the results for cylindrical flanges produced by SPIF with varying tool tip diameters

Precut hole diameter, D_0 (mm)	Tool diameter (mm)	Drawing angle of the intermediate stages, ψ_i (°)					
		65	70	75	80	85	90
102	8						
120							
120	12						
130							
120	25						
130							

Light and dark gray cells correspond to successful and unsuccessful results, respectively.

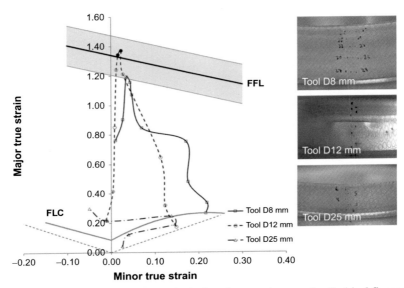

Figure 2.8 Strain loading paths in the principal strain space in case of cylindrical flanged parts made from blanks with a precut hole diameter $D_0 = 120\,\text{mm}$ and various tool tip diameters.

region in Figure 2.9), and four sides, each of them similar to a straight flange produced by incremental bending (refer to the dark gray region in Figure 2.9).

Four different types of meridional sections aa′, bb′, cc,′ and dd′ located at the middle of the side, at the transition between the side and the corner, at the corner, and at the middle of the corner are investigated. The corresponding grid points placed along each of these meridional sections during the initial, intermediate, and final stages of deformation are also identified in Figure 2.9.

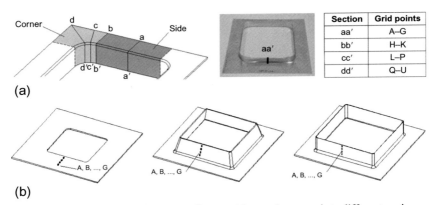

Figure 2.9 (a) Breaking down the square flange with round corners into different regions for analyzing the deformation mechanics of the hole-flanging process and (b) schematic representation of the grid points placed along the middle of a side.

Figure 2.10 was built by tracing the strain paths of the selected grid points "A"–"U" at each intermediate stage of deformation (ψ_i=65–90°) of a square flange with round corners produced by SPIF from a blank with a precut hole with a side length $L_0 = 140$ mm and a corner radius $R_0 = 22.5$ mm.

In case of the section aa' located at the middle of the side, the strain paths of the grid points A–G at the six intermediate drawing angles (ψ_i=65–90°) are plotted as vertical lines under plane-strain conditions (Figure 2.10a). The grid point D placed between the corner radius and the hole edge experiences the highest strain but its overall reduction in thickness $t_{red} = 0.27$ is far below the limiting conditions of failure by fracture given by the FFL. The strain paths of the grid points located at the middle of the corner (section dd') exhibit linear strain paths (refer to the lines-of-best-fit through experimental data in Figure 2.10d) radiating from the origin toward biaxial strain conditions. The maximum strain occurs at point R located 4 mm away from the corner radius in the undeformed blank and corresponds to a thickness reduction $t_{red} = 0.52$.

From what was said above, it is possible to conclude that interaction of corners on the sides and vice versa along meridional sections aa' and dd' is negligible in case of a square flange with round corners produced by SPIF from a blank with a precut hole geometry determined by $L_0 = 140$ mm and $R_0 = 22.5$ mm.

In contrast, the strain envelopes of the grid points belonging to sections bb' and cc' reveal a significant interaction between the corners and the sides. For example, the strain envelope of section cc' (Figure 2.10c) is similar to that previously observed in the hole-flanging of cylinder flanges with a large radius produced by SPIF, because the largest strain value (corresponding to $t_{red} = 0.61$) is located at the middle of the wall undergoing near plane strain conditions and the smallest strain is located at the hole edge under biaxial stretching conditions (refer also to Figure 2.6).

Finally, the strain envelope of section bb' (Figure 2.10b) combines biaxial strain paths that are likely to develop in the corners with radically different, nonlinear strain paths at the hole edge that evolve from near plane strain conditions into uniaxial tension as deformation progresses.

Figure 2.11 provides insight into the physics behind the occurrence of failure when attempting to produce a square flange from a blank with a precut hole with a side length $L_0 = 147$ mm and a very small corner radius of $R_0 = 5$ mm. The flange fails by tearing at the corner with a crack that starts out at the edge and moves down toward the bottom.

The strain envelopes plotted in Figure 2.11 combined with the schematic representation of the deformed circle grid included in Figure 2.11c shows progressive rotation of the deformed circles from near plane strain to uniaxial tension as the grid points move along section cc' toward the hole edge. This result, combined with the biaxial stretching conditions that are available at the middle of the corner (section dd'), produces a significant rise of the strain envelope of section dd' so that the highest strains cross the FFL and lead to failure by fracture.

The morphology of the crack pictured in Figure 2.11b and its propagation path along the meridional direction allows concluding that failure is triggered after exceeding the maximum load-carrying capacity due to the circumferential stresses acting on the corner of the flange. Failure is caused by a significant amount of sheet thinning

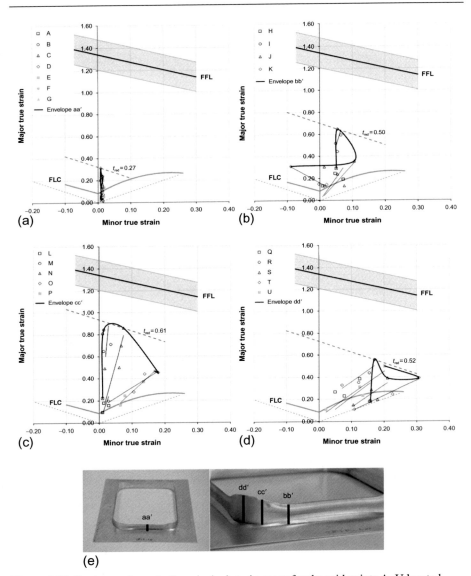

Figure 2.10 Strain envelopes in the principal strain space for the grid points A–U located along different meridional sections (a) aa′, (b) bb′, (c) cc′, and (d) dd′ of a square flange with round corners produced by SPIF from a blank with a precut hole with a side length $L_0 = 140\,\text{mm}$ and a corner radius $R_0 = 22.5\,\text{mm}$ at the drawing angles $\psi_i = 65$–$90°$. The photographs in (e) show the different meridional sections.

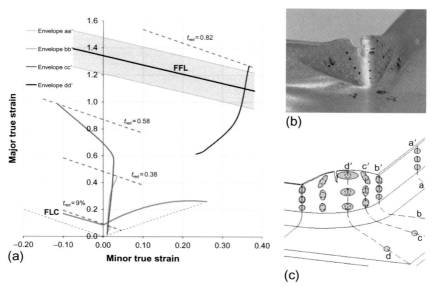

(a)   (b)   (c)

Figure 2.11 (a) Strain envelopes in the principal strain space, (b) photograph showing failure by tearing at the corner, and (c) schematic representation of the deformed circles of the grid at the corner of a square flange with round corners produced by SPIF from a blank with a precut hole with a side length $L_0 = 147\,\text{mm}$ and a corner radius $R_0 = 5\,\text{mm}$ at drawing angles $\psi_i = 65\text{–}80°$.

until fracture (corresponding to a thickness reduction $t_{\text{red}} = 0.82$), without signs of previous localized necking.

2.4.2.2 Process window

Table 2.8 provides the process window as a function of the geometric features of the precut holes. Considering, for example, the influence of the initial side length L_0 of the precut hole on the overall formability of the process, one may conclude that it is not possible to fabricate a square flange with a side length $L = 170$ mm and round corners from blanks with $L_0 < 136$ mm due to the occurrence of failure. In what concerns the initial corner radius R_0 of the precut holes, results show that a larger radius increases the overall formability of the process. In fact, the maximum admissible angle attained in multistage SPIF of a square flange with round corners produced from a blank with a precut hole with $L_0 = 136$ mm was found to decrease from $\psi_6 = 90°$ when $R_0 = 31.5$ mm (where vertical wall flanges are successfully formed) to values of $\psi_4 = 80°$ when $R_0 = 22.5$ mm and $\psi_1 < 65°$ when $R_0 \leq 10$ mm.

The reduction in formability derived from diminishing the side length L_0 and the corner radius R_0 is primarily attributed to the reduction of the initial area $A_0 = L_0^2 - R_0^2 (4 - \pi)$ of the precut holes. This is because small precut holes, experiencing little or no expansion at all (as in the case of $L_0 = 126$ mm and $R_0 = 5$ mm),

Table 2.8 Summary of the results obtained for square flanges with round corners produced by SPIF

Precut hole geometry		Drawing angle of the intermediate stages, ψ_i (°)					
Side length, L_0 (mm)	Corner radius, R_0 (mm)	65	70	75	80	85	90
126	5						
136							
140							
147							
154							
126	10						
136							
140							
147							
154							
126	22.5						
136							
140							
147							
154							
126	31.5						
136							
140							
147							
154							

Light and dark gray cells correspond to successful and unsuccessful results, respectively.

are responsible for approaching plastic flow to that of conventional SPIF of a rectangular pyramid, which fails by cracking at $\psi_{max} < 75°$ (for a reduction in thickness $t_{red} = 0.74$) according to previous experiments performed by Silva et al. (2011).

2.4.3 Complex flanges

Table 2.9 provides the process window of complex flanges combining features of cylindrical and square flanges as a function of the geometric features of the precut holes. As seen, it was not possible to fabricate complex flanges with vertical walls due to the occurrence of failure when the maximum drawing angle of the intermediate stage was $\psi_i \geq 80°$.

Experimental observations of four different types of meridional cross sections: (i) aa′ located at the middle of the round side, (ii) bb′ and dd′ located at the middle of the straight sides, and (iii) cc′ located at the middle of the corner (Figure 2.12) revealed two different types of failure. When $R_0 < 20\,mm$, flanges fail by tearing at the corner similarly to what was previously observed in hole-flanging of square flanges with very

Table 2.9 **Summary of the results obtained for complex flanges produced by SPIF**

Precut hole geometry		Drawing angle of the intermediate stages, ψ_i (°)						
D_0 (mm)	R_0 (mm)	65	70	75	80	85	90	
128	10							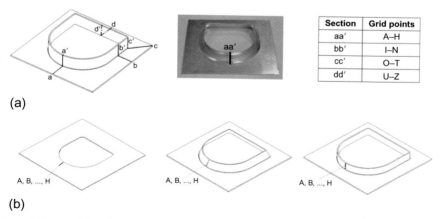
	15							$R_0 = 10$ mm
	20							
	30							

Light and dark gray cells correspond to successful and unsuccessful results, respectively.

small corner radii (refer to the picture in Table 2.9). In contrast, when $R_0 \geq 20$ mm, flanges fail by cracking at the transition between the corners and the straight sides.

The strain envelopes associated with failure by tearing at the corner when $R_0 = 10$ mm are shown in Figure 2.13. As seen in the figure, the envelopes of the cross section cc′ cross the FFL in close agreement with the experimental observation of failure by tearing at the corner.

The final conclusion arising from this preliminary research on complex flanges produced by SPIF is that geometric features of the precut holes must favor straight and round sides with a large radius instead of corners with very small radii. A small corner radius, as in the case of square flanges, will lead to failure by tearing with cracks that move toward the corner edges.

Section	Grid points
aa′	A–H
bb′	I–N
cc′	O–T
dd′	U–Z

(a)

(b)

Figure 2.12 Breaking down the complex flange into different regions for analyzing the deformation mechanics of the hole-flanging process.

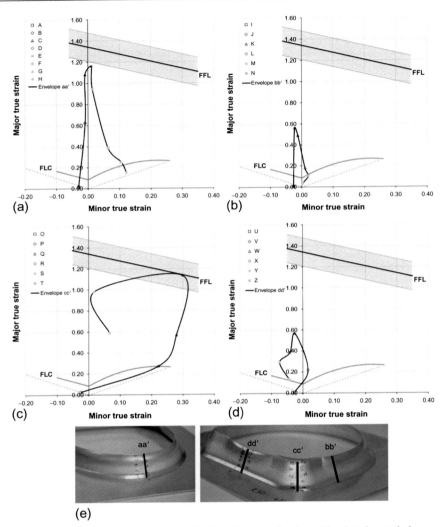

(e)

Figure 2.13 Strain envelopes in the principal strain space for the grid points located along different meridional sections (a) aa′, (b) bb′, (c) cc′, and (d) dd′ of a complex flange produced by SPIF from a blank with a precut hole with $D_0 = 128\,mm$ and $R_0 = 10\,mm$ at the drawing angles $\psi_3 = 75°$. The photographs in (e) show the different meridional sections.

2.5 Conclusions

The morphology of the cracks and the propagation paths along the circumferential direction allow concluding that cylindrical flanges produced by SPIF fail by fracture with suppression of necking in crack opening mode I. In general, the minimum precut hole diameter of the blank increases as the tool diameter increases in close agreement with the well-known reduction of formability with increasing tool diameter that is commonly observed in conventional SPIF (Silva et al., 2009).

In case of hole-flanging produced by SPIF being performed with forming tools having very large diameters, plastic flow and failure approach the operating conditions of hole-flanging produced by conventional press-working and give rise to a reduction of the total height of the flange with the increase of the tool diameter.

In the case of square and complex flanges, the interaction between the forming tool and the blank at different locations of the hole-flanged parts gives rise to significant differences in plastic flow and failure. Square flanges with very small corner radii, for example, give rise to crack morphology and propagation paths that allow characterizing failure to be triggered at the corner edges by circumferential stresses after exceeding the maximum load-carrying capacity of the blanks and moving down along the meridional direction.

Failure takes place by thinning until fracture without signs of previous localized necking as in the case of cylindrical flanges and may be prevented by changing the geometry of the precut holes by increasing the side length L_0, the corner radius R_0, or both simultaneously.

In case of complex flanges, small corners were found to give rise to failure by tearing in close agreement to that observed in hole-flanging of square flanges produced by SPIF. Larger corners, but still below the values that are needed to successfully produce a sound complex flange, also lead to failure by cracking at the transition between the corners and the straight sides. The design of complex hole-flanged parts produced by SPIF must favor straight and round sides with large radii instead of corners with very small radii.

Acknowledgments

The work of André Teodora, Carlos Silva, Jacob Torstein Henckel, João Soeiro, Mikkel Ravn Boye Jensen, and Pedro Pardal is gratefully acknowledged. P.A.F. Martins and M.B. Silva would like to acknowledge the support provided by the Portuguese Foundation for Science and Technology.

References

ASTM Standard E8/E8M *Standard test methods for tension testing of metallic materials.* (2013). West Conshohocken, PA: ASTM International.

Atkins, A. G. (1996). Fracture in forming. *Journal of Materials Processing Technology, 56,* 609–618.

Centeno, G., Silva, M. B., Cristino, V. A. M., Vallellano, C., & Martins, P. A. F. (2012). Hole-flanging by incremental sheet forming. *International Journal of Machine Tools and Manufacture, 59,* 46–54.

Cristino, V. A. M., Montanari, L., Silva, M. B., & Martins, P. A. F. (2014). Towards square hole-flanging produced by single point incremental forming. *Journal of Materials: Design and Applications,* http://dx.doi.org/10.1177/1464420714524930.

Cui, Z., & Gao, L. (2010). Studies on hole-flanging process using multistage incremental forming. *CIRP Journal of Manufacturing Science and Technology, 2,* 124–128.

Glover, G., Duncan, J. L., & Embury, J. D. (1977). Failure maps for sheet metal. *Metals Technology, 4,* 153–159.

Hill, R. (1948). A theory of yielding and plastic flow of anisotropic metals. *Proceedings of the Royal Society of London (Series A), 193,* 281–297.

Hirt, G., Ames, J., Bambach, M., & Kopp, R. (2004). Forming strategies and process modelling for CNC incremental sheet forming. *CIRP Annals—Manufacturing Technology*, *53*, 203–206.

Isik, K., Silva, M. B., Tekkaya, A. E., & Martins, P. A. F. (2014). Formability limits by fracture in sheet metal forming. *Journal of Materials Processing Technology*, *214*, 1557–1565.

Jeswiet, J., Micari, F., Hirt, G., Bramley, A., Duflou, J., & Allwood, J. (2005). Asymmetric single point incremental forming of sheet metal. *Annals of CIRP*, *54*, 623–650.

Johnson, W., Chitkara, N. R., & Minh, H. V. (1977). Deformation modes and lip fracture during hole-flanging of circular plates of anisotropic materials. *Journal of Engineering for Industry, Transactions of ASME*, *99*, 738–748.

Lang, L. H., Wang, Z. R., Kang, D. C., Yuan, S. J., Zhang, S. H., Danckert, J., et al. (2004). Hydroforming highlights: Sheet hydroforming and tube hydroforming. *Journal of Materials Processing Technology*, *151*, 165–177.

Mackerle, J. (2004). Finite element analyses and simulations of sheet metal forming processes. *Engineering Computations*, *21*, 891–940.

Martins, P. A. F., Bay, N., Tekkaya, A. E., & Atkins, A. G. (2014). Characterization of fracture loci in metal forming. *International Journal of Mechanical Sciences*, *83*, 112–123.

Montanari, L., Cristino, V. A. M., Silva, M. B., & Martins, P. A. F. (2013). A new approach for deformation history of material elements in hole-flanging produced by single point incremental forming. *International Journal of Advanced Manufacturing Technology*, *69*, 1175–1183.

Muscat-Fenech, C.M., Arndt, S., & Atkins, A.G., 1996. *The determination of fracture forming limit diagrams from fracture toughness. Proceedings of the 4th international sheet metal conference,* Vol. 1 (pp. 249–260). The Netherlands: University of Twente.

Nimbalkar, D. H., & Nandedkar, V. M. (2013). Review of incremental forming of sheet metal components. *International Journal of Engineering Research and Application*, *3*, 39–51.

Park, Y., & Colton, J. S. (2003). Sheet metal forming using polymer composite rapid prototype tooling. *Journal of Engineering Materials and Technology*, *125*, 247–255.

Petek, A., Kuzman, K., & Fijavž, R. (2011). Backward drawing of necks using incremental approach. *Key Engineering Materials*, *473*, 105–112.

Silva, M. B., Nielsen, P. S., Bay, N., & Martins, P. A. F. (2011). Failure mechanisms in single point incremental forming of metals. *International Journal of Advanced Manufacturing Technology*, *56*, 893–903.

Silva, M. B., Skjoedt, M., Atkins, A. G., Bay, N., & Martins, P. A. F. (2008). Single point incremental forming and formability/failure diagrams. *Journal Strain Analysis in Engineering Design*, *43*, 15–36.

Silva, M. B., Skjoedt, M., Bay, N., & Martins, P. A. F. (2009). Revisiting single-point incremental forming and formability/failure diagrams by means of finite elements and experimentation. *Journal of Strain Analysis*, *44*, 221–234.

Silva, M. B., Teixeira, P., Reis, A., & Martins, P. A. F. (2013). On the formability of hole-flanging by incremental sheet forming. *Journal of Materials Design and Applications*, *227*, 91–99.

Stachowicz, F. (2008). Estimation of hole-flange ability for deep drawing steel sheets. *Archives of Civil and Mechanical Engineering*, *8*, 167–172.

Voswinckel, H., Bambach, M., & Hirt, G. (2013). Process limits of stretch and shrink flanging by incremental sheet metal forming. *Key Engineering Materials*, *549*, 45–52.

Wang, N. M., & Wenner, M. L. (1974). An analytical and experimental study of stretch flanging. *International Journal of Mechanical Sciences*, *16*, 135–143.

Wong, C. C., Dean, T. A., & Lin, J. (2003). A review of spinning, shear forming and flow forming processes. *International Journal of Machine Tools and Manufacture*, *43*, 1419–1435.

Yamada, Y., & Koide, M. (1968). Analysis of the bore-expanding test by the incremental theory of plasticity. *International Journal of Mechanical Sciences*, *10*, 1–14.

Flexible roll forming

3

M.M. Kasaei[1], H. Moslemi Naeini[1], B. Abbaszadeh[1], M.B. Silva[2],
P.A.F. Martins[2]
[1]Tarbiat Modares University, Tehran, Islamic Republic of Iran; [2]Universidade de Lisboa,
Lisboa, Portugal

3.1 Introduction

Roll forming is a metal forming process in which a long strip of sheet metal is continuously bent in stages by passing it through a series of contoured driven rolls placed in different stands along a line. The industries which make use of the roll forming process are very diverse but typical products consist of cross section profiles with long lengths manufactured in large quantities at high production rates.

The fundamentals of conventional roll forming are available in textbooks of metal forming, but the principles behind roll form tool design are not so widely accessible in that only a few references exist in the open technical literature (Alvarez, 2006).

Unlike conventional roll forming, the rolls of the flexible roll forming (FRF) process are not fixed in their position and can be moved along a path which describes the desired bend line of the profile. The translational and rotational movement of the rolls is controlled by computerized numerical control and their position is such that they are always tangent to the deformed flange to avoid additional, undesirable, and plastic deformation (Figure 3.1). FRF allows producing variable cross section profiles for the automotive, railway, ship construction, and building industries, among others.

The pioneering developments in the field were carried out by the Swedish company ORTIC AB in 2001. The company developed a 3D roll forming machine to produce longitudinally curved and variable width panels for covering roofs of buildings worldwide. The design of the machine entailed letting the steel strip pass straight through it while cutters and roll form tools move along a tapered path.

Groche, von Breitenbach, Jckel, and Zettler (2003) presented a new tooling concept in which a single flexible stand was integrated in a conventional roll forming line for producing profiles with different lengths and variable cross sections. They showed how rolls must be moved to avoid undesirable deformation in the variable cross section profiles by means of experimental tests and numerical simulations. This variant of roll forming was named flexible roll forming and its utilization is generally combined with laser cutting or roll slitting to produce the tailored precut strips that need to be fed into the machine.

In a subsequent publication, Groche, Zettler, and Berner (2006) investigated the finite element distribution of strain along the flanges of the profiles and proposed an analytical "one-step-model" to predict the occurrence of wrinkling and to allow design and selection of process parameters in FRF. The model is one-dimensional and takes into consideration the fundamentals of the plastic instability of plates, the

Materials Forming and Machining. http://dx.doi.org/10.1016/B978-0-85709-483-4.00003-X

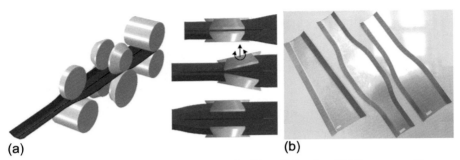

(a)   **(b)**

Figure 3.1 (a) Schematic representation of the flexible roll forming (FRF) process with a detail showing the forming stand rolls changing position during the process. (b) Examples of channel profiles produced by FRF at the Tarbiat Modares University (Iran).

strains acting in the compression zone of the flanges, and the interaction between the stretched and compressed zones of the flanges by means of correction factors derived from finite element computations.

Ona (2005) introduced an intelligent FRF machine with a single forming stand that allows rotating, turning, and moving the rolls in and out for producing profiles with variable cross sections. Their work also includes a comprehensive experimental investigation on the major defects that are likely to occur on the web (distortion/warping) and flange regions (wrinkling) of the variable cross section profiles. Later, Ona, Sho, Nagamachi, and Hoshi (2010) developed a tandem FRF machine based on their previous concept and proposed guidelines to eliminate wrinkling and web warping that were aimed at reducing the longitudinal strain at the compression zone and increasing the longitudinal strain at the stretching zone, respectively.

Lindgren (2009) developed a 3D roll forming line for producing hat profiles and showed its feasibility to produce variable cross section hat profiles in both width and depth directions with tolerances similar to those of straight profiles.

Recent applications of the finite element method to simulate FRF proved capable of predicting shape defects and the distribution of the major field variables. Gülçeken, Abeé, Sedlmaier, Livatyali (2007), for example, set up a finite element model to investigate the feasibility of the process and to determine the regions of the flanges where compressive stresses are likely to develop. Abee, Berner, and Sedlmaier (2008) performed a finite element-based investigation on a blank holder system with transverse movement in front of the FRF stand to support the web of the profile and concluded that such a system would lead to significant improvements in the overall accuracy of profiles.

Larrañaga and Galdos (2009) and Larrañaga, Galdos, García, Ortubay, and Arrizabalaga (2008) made use of the finite element method to investigate the feasibility of applying laser assisted heating to improve formability and decrease the geometrical deviation in the FRF of high strength steels. They investigated different heating strategies by using numerical simulation and identified the vertical parts of stretching and compression zones in the hat profiles as the best regions for local heating.

Vogler, Duschka, and Groche (2010) made use of finite elements to investigate the possibility of producing closed profiles with variable cross sections and concluded that a combination of FRF and conventional roll forming could be successfully applied to that purpose. Yan and Li (2011) simulated the nine-step FRF of an ultra-high-strength steel bumper to determine the occurrence of forming defects and to estimate the forming load and torque of the rolls due to its importance in process design. Zhao, Yan, Wang, and Gao (2013) performed finite element modeling of FRF to analyze the evolution of ductile damage.

Gao, Li, and Zhao (2013) enhanced previous thermal–mechanical finite element analysis of FRF by using a sequence-based approach to determine the effect of local heating in FRF. Hennig, Sedlmaier, and Abée (2011) presented a comprehensive review of published finite element simulations of FRF along the timeline of development of the process and discussed the utilization of numerical simulations to investigate the overall feasibility of the process and roll design.

Park, Yang, Cha, Kim, and Nam (2014) combined finite element analysis and experimentation to develop a variant of FRF named incremental counter forming (ICF) that is aimed at reducing the shape error in the FRF process by increasing the longitudinal strain at the stretching zone. Three stands of ICF rolls were placed between the FRF stands in contact with the web section of the profile so that longitudinal strain distribution at the stretching zone could be controlled by adjusting the major forming parameters (the amount of counter forming, the span length between the forming rolls and the ICF roll set, and the transverse gap between the ICF roll sets).

Very recently, Kasaei et al. (2014) and Yan, Wang, Li, Qian, and Mpofu (2014) set up finite element models to determine the distribution and time history of strain in both constant and variable cross sections of the profiles. Both publications are a major step forward in the understanding of the forming mechanisms and deformation mechanics of FRF but the distribution and time history of stresses are not covered with the same level of detail as that of strains. Moreover, in the case of Yan et al. (2014), the validation of the overall numerical simulations was performed by measuring the geometry of the formed cross sections and comparing experimental shapes with those predicted by finite elements.

The aims and objectives of this chapter are in line with those of Kasaei et al. (2014) and Yan et al. (2014) but the overall approach is different, with stronger emphasis on the combined distribution and evolution of stresses and strains with time. As a result of this, a new formability analysis based on the evolution of effective strain vs. stress triaxiality along transverse and longitudinal directions is proposed to predict and explain the occurrence of flange wrinkling in variable cross section profiles produced by FRF. Experiments with different profile shapes give support to the presentation and measurements of longitudinal strain history at the flange edge using resistance strain gauge technology (Kasaei et al., 2014) are used to evaluate the overall accuracy and reliability of finite element predictions.

3.2 Experimentation

3.2.1 Material characterization

The research work was carried out on St12 steel sheets with 0.5 mm thickness. The mechanical characterization of the material was performed by means of tensile tests in an Instron 5500R testing machine with specimens that were cut out from the supplied sheets at 0°, 45°, and 90° with respect to the rolling direction. The resulting average stress–strain curve was approximated by the following Ludwik–Hollomon's equation:

$$\sigma = 586.8\varepsilon^{0.246}\,(\text{MPa}) \tag{3.1}$$

The values obtained for the modulus of elasticity E, the yield strength σ_Y, the ultimate tensile strength σ_UTS, the anisotropy coefficient r, and the elongation at break A at 0°, 45°, and 90° with respect to the rolling direction are provided in Table 3.1. The normal $\bar{r}$ and planar Δr anisotropy coefficients were determined from

$$\bar{r} = \frac{r_0 + 2r_{45} + r_{90}}{4}, \quad \Delta r = \frac{r_0 - 2r_{45} + r_{90}}{2} \tag{3.2}$$

3.2.2   Methods and procedures

The FRF equipment utilized for the production of profiles with variable cross sections was designed, fabricated, and instrumented at the Tarbiat Modares University. For purposes of presentation, and although not corresponding to what readers directly observe in Figure 3.2, the equipment will be split into three major parts:

(i)   Forming stand
(ii) Feeding mechanism
(iii) Control unit

The forming stand consists of two pairs of rolls that are integrated in a parallel kinematic system that allows following the varying contours of the profiles and distributing the applied loads on each side along two different axes. Because the setup was developed to produce symmetric profiles, servo motors are applied at one side of the forming stand and movements are transferred symmetrically to the other side by connecting the right hand thread ball screw to the left hand thread ball screw along an axis. The housings are installed on a carriage that is mounted on linear guide rails placed along each side of the forming stand. The carriage contains bearings with a special arrangement for allowing the rotational movement of the housings. The rotational movement is achieved by allowing the drives to work in opposite directions. The translational movement of the housings is accomplished by operating both drives in the same direction. Combination of the rotational and translational movements allows the rolls to follow the required forming contours of the variable cross section profiles.

The feeding mechanism pulls the sheet into the rolls and provides the main driving force during forming. This is accomplished by means of a fixture mounted on two linear guides with translational movement and it is necessary because one forming stand would not be sufficient to ensure precise feeding of the sheet by friction force if the rotating rolls were driven. At the beginning of the roll forming process, the two pairs of rolls are positioned into a channel type configuration (refer to Figure 3.2) to allow the initial part of the strip to be fed into the rolls. Then, translational and rotational movements of the housing allow the rolls to impose the required geometry of the profile while the strip is being fed.

Table 3.1 **Summary of the mechanical properties of the St12 steel sheets that were utilized in the investigation**

	Modulus of elasticity, E (GPa)	Yield strength, σ_Y (MPa)	Ultimate tensile strength, σ_{UTS} (MPa)	Anisotropy coefficient, r			Normal anisotropy, $\bar{r}$	Planar anisotropy, Δr	Elongation at break, A (%)
				0	45	90			
St12	196.53	173.27	327.68	1.32	1.23	1.55	1.30	0.20	55.26

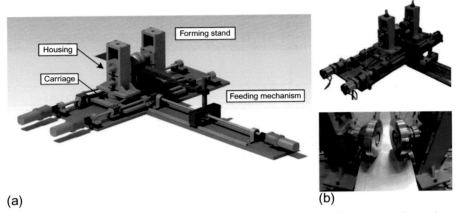

(a)   (b)

Figure 3.2 (a) The FRF setup that was utilized in the investigation with main notation and (b) pictures showing the experimental apparatus and a detail of the forming stand.

The control unit is built on a personal computer equipped with a data acquisition board (ADVANTECH-PCI1220U) and a Labview-based software that allows users to input the geometry and velocity of the profiles in order to transform these inputs into the corresponding movements of the housing and feeding mechanism.

The experimental work plan was designed in order to produce variable cross section profiles with different geometric parameters. Figure 3.3 shows schematic drawings of a precut strip and a variable cross section profile with indication of the following geometric parameters:

(i)   Radius R of the bending line in the stretching and compression zones
(ii) Radius of curvature R_1 at the edge of the stretching zone
(iii) Radius of curvature R_2 at the edge of the compression zone
(iv) Width W_t of the transition zone
(v)   Widths W_1 and W_2 of the initial and final webs
(vi) Flange length F

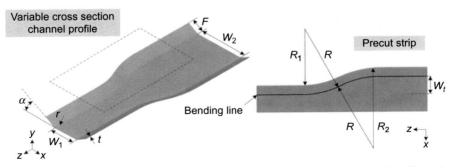

Figure 3.3 Schematic geometry with notation of a variable cross section channel profile and its precut strip.

(vii) Forming angle α
(viii) Strip thickness t
(ix) Bend radius r at the transition between the flange and the web

The experimental work plan is summarized in Table 3.2.

Unlike other sheet metal forming processes where the experimental in-plane strains can be obtained from circle grid analysis, in FRF this is not feasible because the strains are very small and highly localized. As a result of this, the authors employed resistance strain gauge technology to register and evaluate the history of longitudinal strain in specific locations. A strain meter TML DC-97A and strain gauges TML FLA 5–11 were utilized (Figure 3.4a). The strain gauges were mounted at the middle of the stretching and compression top surface zones along the longitudinal direction and 2.5 mm away from the edge. A special roll with a flange length 10 mm smaller than that of the profile was employed in order to prevent strain gauges from being damaged (refer to the rectangle in Figure 3.4b).

The web centerline profile at transition zone was measured after the forming process in order to evaluate its geometric accuracy. For this purpose, start and end points of the transition zone were located at the same height related to a granite surface plate by using gauge blocks and then a dial indicator was applied for measuring height of 20 points along the web centerline (Figure 3.5).

3.3   Finite element modeling

The finite element modeling of FRF was performed in the commercial computer program ABAQUS. A quasistatic implicit formulation was utilized because the velocity of the strip was small and because alternative dynamic explicit formulations are known to be less accurate and more sensitive to input data related to mesh discretization and kinematic operating parameters (Tekkaya & Martins, 2009).

Elasto-plastic constitutive equations were chosen in order to properly account for springback. Isotropic work-hardening was considered and the stress–strain response of the material in the plastic domain was approximated by means of a Ludwik–Hollomon's equation (refer to Equation (3.1)). The forming rolls were modeled as analytical shell rigid bodies because their deformation is negligible during the entire process (Figure 3.6a). The strips were discretized by means of four-node shell elements ("S4R" type in ABAQUS) with five integration points through the thickness (Figure 3.6b). Computations made use of reduced integration procedures and a penalty-based approach was utilized for modeling the contact with friction between the rolls and strip. The Coulomb friction law was employed and a friction coefficient $\mu = 0.1$ was assumed.

Because the profiles and forming stand are symmetric along the longitudinal centerline, only half of the strip and rolls needed to be modeled (Figure 3.6b). The strip was pulled longitudinally into the forming stand by means of boundary conditions consisting of a constant velocity equal to 20 mm/s applied in the leading edge. Translational and rotational movements of the housing were calculated and defined for each forming roll in order to replicate the geometrical features of the variable cross section profiles.

Table 3.2 The experimental work plan

Test case	R (mm)	R_1 (mm)	R_2 (mm)	W_t (mm)	W_1 (mm)	F (mm)	α (°)	t (mm)	Location of strain gauge
1	400	370	430	35	70	30	20	0.5	Stretching zone
2	400	370	430	35	70	30	20	0.5	Compression zone
3	400	370	430	35	70	30	30	0.5	Stretching zone
4	400	370	430	35	70	30	30	0.5	Compression zone
5	400	370	430	35	70	30	40	0.5	Stretching zone
6	400	370	430	35	70	30	40	0.5	Compression zone

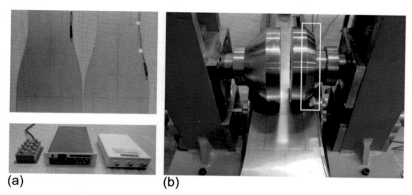

Figure 3.4 (a) Strain gauges and equipment utilized for measuring the longitudinal strains. (b) Forming stand with a special top roll for preventing strain gauges from being damaged.

Figure 3.5 Method utilized for measuring the web centerline profile.

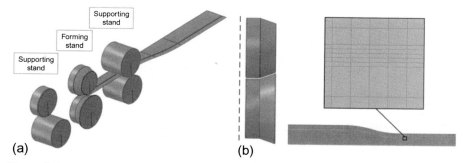

Figure 3.6 (a) Schematic view of the FRF numerical model. (b) Detail of a roll and strip in the forming stand showing symmetry and discretization by shell elements.

Table 3.3 **The work plan for the numerical modeling based on Strategy 2**

Test case	R (mm)	R_1 (mm)	R_2 (mm)	W_t (mm)	W_1 (mm)	F (mm)	$\alpha(°)$	t (mm)
1	400	380	420	35	70	20	10	1
2	400	380	420	35	70	20	20	1
3	400	380	420	35	70	20	30	1

Finite element modeling of the FRF process was accomplished by means of two different strategies hereafter designated as "Strategy 1" and "Strategy 2." Strategy 1 is directly connected to the experimental work plan that was previously described in Section 3.2.2 and was employed with the purpose of validating the finite element models against experimental data and observations.

The aim and objectives of Strategy 2 were different and focused on removing the influence of web warping on flange wrinkling. This was accomplished by fixing the displacement of the web in the strip thickness direction. The flange length F and the minimum forming angle α were smaller than in the case of Strategy 1, with values, respectively, equal to 20 mm and 10°. The work plan for the numerical modeling based on Strategy 2 is summarized in Table 3.3.

The stress–strain response of the material utilized in Strategy 2 was also characterized by means of a Ludwik–Hollomon's equation but instead of using an equation similar to that of the experiments (as in the case of Strategy 1), it was decided to allow several of its parameters to vary in order to investigate their influence in the overall performance of FRF. Anisotropy was not taken into consideration in the numerical simulations performed with Strategy 2.

Typical finite element meshes resulting from the two above-mentioned simulation strategies were made of approximately 7800 nodal points and 7500 four-node shell elements. The overall CPU time on a standard personal computer equipped with an Intel i7 CPU (3.5 GHz) processor was below 24 h for Strategy 1 and below 18 h for Strategy 2 due to differences in convergence stability resulting from fixing the displacement of the web in the strip thickness direction.

3.4   Results and discussion

3.4.1   Strain distribution and web centerline profile

Figure 3.7 presents a comparison between experimental and numerical predictions of longitudinal strain and web centerline profile. The numerical predictions were obtained from the modeling approach previously designated as Strategy 1 (refer to Section 3.3). The values of the longitudinal strains were taken from the middle of the stretching and compression zones whereas the values of the web centerline profile were taken from the transition zone.

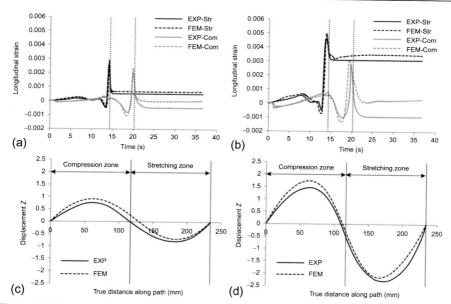

Figure 3.7 Experimental and finite element predicted evolution of longitudinal strain with time for (a) test cases 1 and 2 and (b) test cases 5 and 6 of Table 3.1. Vertical displacement of the web centerline along longitudinal direction for (c) test cases 1 and 2 and (d) test cases 5 and 6 of Table 3.1. *Note*: The abbreviations "Com" and "Str" denote compression zone and stretching zone, respectively.

As seen in Figure 3.7a and b, the level of the peak strains increases with the forming angle α and develops slightly before material crosses the mid-section of the roll in the stretching and compression zones. Results also show final strains in the stretching zone to increase with the increase of the forming angle and minor differences between the experimental and the finite element predicted strains to develop in the compression zone. These differences may be attributed to elastic unloading (springback) and consist of very small values, with limited meaning.

The evolution of finite element predicted strains with time shows a general good agreement with experimental measurements and allows concluding that numerical models utilized in the overall investigation are adequate to simulate FRF.

The web centerline profiles shown in Figure 3.7c and d are typical of FRF without a blank holder and are characterized by a "sinusoidal-type" shape that is originated by redundancy and insufficiency of material along the flanges propagating into the web. This phenomenon leads to undesirable warping defects along the web, which may be diminished by increasing the overall strain level in the flange.

The differences between experimental measurements and numerical predictions are compatible with those observed in Figure 3.7a and b and may also be attributed to elastic unloading.

3.4.2 Physics of FRF

Unlike previous publications, which explain the physics of FRF by means of strain-based methodologies, the focus of this chapter is placed on the combined evolutions of stress and strain with time. This allows explaining the occurrence of flange wrinkling in variable cross section profiles by means of the evolution of the effective strain vs. stress triaxiality along transverse and longitudinal directions.

Figure 3.8 shows the evolution of strain and stress with time obtained from finite element analysis with test cases 1 and 3 of Strategy 2 (refer to Table 3.3) for selected locations at the middle of the bending area in the compression zone of the variable cross section profiles.

As seen from the finite element predicted evolution of strains with time (Figure 3.8a and c), plastic deformation in the bending area approaches plane strain conditions,

$$d\varepsilon_{22} = \frac{d\bar{\varepsilon}}{\bar{\sigma}} \left\{ \sigma_{22} - \frac{1}{2}(\sigma_{11} + \sigma_{33}) \right\} = 0 \tag{3.3}$$

meaning that material flow in the bending area is always parallel to plane "1–3" (refer to the coordinate system included in Figure 3.8). Because there are no shear stresses acting in the bending area, the stress σ_{22} acting along the longitudinal direction is a principal stress and its value can be calculated from

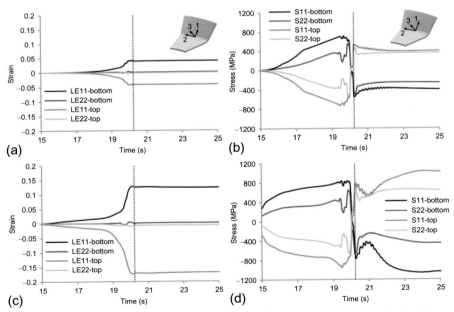

Figure 3.8 Finite element results in the bending area of the compression zone obtained by means of Strategy 2 (Table 3.3). (a) In-plane strains vs. time for case 1, (b) in-plane stresses vs. time for case 1, (c) in-plane strains vs. time for case 3, and (d) in-plane stresses vs. time for case 3. *Note*: The black vertical line represents the instant of time when material crosses the mid-section of the roll.

$$\sigma_{22} = \frac{1}{2}\left(\sigma_{11} + \sigma_{33}\right) \tag{3.4}$$

If plane stress conditions $\sigma_{33} = 0$ are further assumed in the bending area, it follows that

$$\sigma_{22} = \frac{\sigma_{11}}{2} \tag{3.5}$$

This result is in close agreement with the finite element predicted evolution of stresses with time that are shown in Figure 3.8b and d, which show that bottom fibers at the bending area are subjected to tension whereas top fibers are subjected to compression. The main difference between the evolutions of stress with time that are shown in Figure 3.8b and d is related to the forming angle α because stresses are expected to increase with the forming angle as a result of the decrease in the bend radius r at the transition between the flange and the web.

After crossing the mid-section of the roll, material starts unloading and the fibers that were carrying compression become subjected to tension and those which were carrying tension become subjected to compression. Elastic unloading leads to the residual stresses that are plotted in the rightmost part of the evolutions of stresses with time in Figure 3.8b and d and is schematically illustrated in Figure 3.9.

Figure 3.10 illustrates the above-mentioned change of sign between loading and unloading by means of the corresponding Mohr stress circles.

A similar analysis could be performed for selected locations placed at the middle of the bending area of the stretching zone but the overall conclusions would be identical to those of the compression zone because differences between these two zones are only relevant in the flanges of the profiles.

Considering the compression zone, for example, the distribution of strains with time for cases 1 and 3 of Table 3.3 in the flange zone that are provided in Figure 3.11a and c, one can conclude that the bottom and top fibers are subjected to uniaxial loading because the ratio of the in-plane strains is approximately given by $\varepsilon_{11} : \varepsilon_{22} = \pm 0.5 : \mp 1$. This conclusion is compatible with the corresponding distribution of stresses with time that is provided in Figure 3.11b and d, $\sigma_{11} \approx 0, \sigma_{22} \neq 0$.

As with the deformation of the bending area, the deformation of the flange edge can be considered in two sections: loading and unloading. The loading section consists of two subsections which are shown in Figure 3.11e. In the first subsection, the

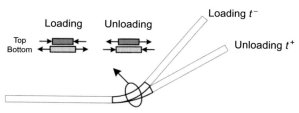

Figure 3.9 Schematic representation of the bending area of the stretching zone at loading "t^-" and unloading "t^+" instants of time.

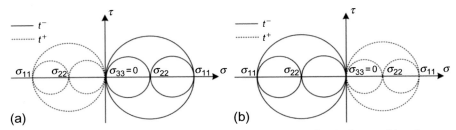

Figure 3.10 Schematic representation of the Mohr stress circles for the bottom (a) and top (b) fibers of the bending area of the stretching zone at loading "t^-" and unloading "t^+" instants of time.

flange edge is not in contact with the roll, and the deformation mechanics may be considered as a combination of bending and membrane stretching so that an increase in the forming angle α will increase membrane deformation more significantly than bending deformation. As a result of this, longitudinal strain in the top surface, which is compressive at the forming angle $\alpha = 10°$, becomes tensile at the forming angle $\alpha = 30°$ and the tensile longitudinal strain in the bottom surface increases with the forming angle.

In the second subsection, flange edge is longitudinally bent by the bottom roll so that longitudinal strain in the bottom and top surfaces is relevant to the radius of the roll. Therefore, an increase of the forming angle has no effect on the longitudinal strain and the difference between the longitudinal strains in the top and bottom surfaces remains constant as illustrated in Figure 3.11a and c. These kinds of deformations were reported in the conventional roll forming process (Azizi Tafti, 2014).

Elastic unloading after material crosses the mid-section of the roll gives rise to changes in the signs of strains and stresses as shown in Figure 3.11e. The strains and stresses developing in the unloading section become the final residual stresses of the variable cross section profile.

Deformation in the unloading section may be considered as stable or unstable. In the case of stable deformation, corresponding to forming angle $\alpha = 10°$, residual strains in the top and bottom surfaces become compressive and approximately identical to each other. In contrast, the forming angle $\alpha = 30°$ experiences signs of unstable deformation leading to differences in residual longitudinal strains between the top and bottom surfaces.

The higher level of compression stresses acting in the bottom and top fibers of case 3 warn of the possibility of failure by wrinkling being more likely to occur when the forming angle α increases from 10° to 30°. However, instead of analyzing the occurrence of wrinkling by means of simplified one-dimensional models based on the plastic instability of plates, the authors will make use of the loading paths that connect strains and stresses in the space of effective strain $\bar{\varepsilon}$ vs. stress triaxiality ratio $\beta = \sigma_m / \bar{\sigma}$ (where σ_m is the mean normal stress and $\bar{\sigma}$ is the effective stress).

Figures 3.12 and 3.13 present such loading paths for the transverse and longitudinal cross sections of the profiles, respectively. In case of Figure 3.12, plots are made for the compression zone at unloading time "t^+" (that is, after crossing the mid-section

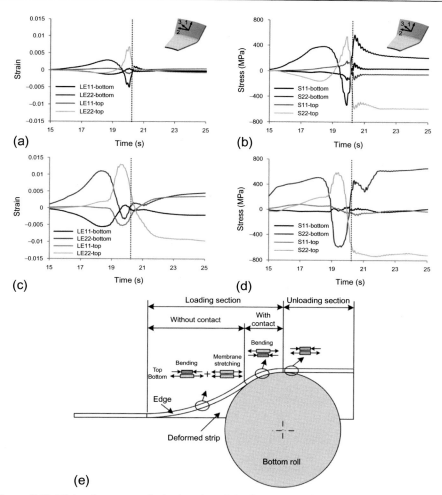

Figure 3.11 Finite element results in the edge of the flange area of the compression zone using Strategy 2 (Table 3.3). (a) In-plane strains vs. time for case 1, (b) in-plane stresses vs. time for case 1, (c) in-plane strains vs. time for case 3, (d) in-plane stresses vs. time for case 3, and (e) schematic representation of material elements in the edge of the flange area of the compression zone during loading and unloading. *Notes*: The black vertical line represents the instant of time when material crosses the mid-section of the roll. For better visualization top roll was removed from this schematic representation.

of the roll), because wrinkling at loading time "*t⁻*" (that is, before crossing the mid-section of the roll) is potentially less critical due to the fact that the bottom surface of the strip is supported by the bottom rolls. The plots in Figure 3.13 are made after forming the profile.

As seen in Figure 3.12, the top fiber of the edges of both test cases are subjected to pure compression $\beta = -1/3$ and the top fibers of the bending areas are subjected

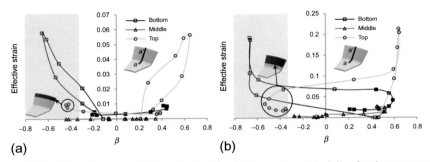

(a)   (b)

Figure 3.12 Finite element evolution of effective strain vs. stress triaxiality for the transverse cross section of the profile after material crosses the mid-section of the roll for (a) case 1 and (b) case 3 of Table 3.3. *Note*: Gray filled markers correspond to the flange area.

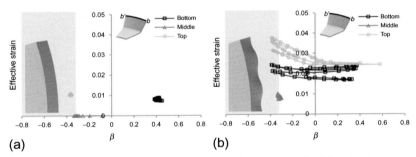

(a)   (b)

Figure 3.13 Finite element evolution of effective strain vs. stress triaxiality for the longitudinal cross section. (a) Case 1 and (b) case 3 after forming the profile.

to $\beta > 0$ after elastic unloading due to material crossing the mid-section of the roll. However, the evolution of the loading paths along the transverse cross section for both test cases allows concluding that there is a larger portion of flange located in the critical region to the left of pure compression $\beta < -1/3$, where wrinkling is likely to occur, in case 3 ($\alpha = 30°$) than in case 1 ($\alpha = 10°$).

The middle and bottom fibers behave differently. The middle fibers experience almost no strain because their location is close to the neutral fibers. The bottom fibers experience near symmetric loading paths to those of the top fibers and, therefore, although a significant amount of material is located to the left of pure compression $\beta < -1/3$, where wrinkling is likely to occur, this will not happen because the top bending radius introduces additional stiffness.

This conclusion that the space of effective strain $\bar{\varepsilon}$ vs. stress triaxiality ratio $\beta = \sigma_m / \bar{\sigma}$ can be used to investigate the risk of wrinkling is further supported by observation of Figure 3.13 where the loading paths along the longitudinal cross section present two completely different trends. Case 1 presents two large spots corresponding to the top fibers under compression and the bottom fibers under tension. In contrast, case 3 presents "wave-type" loading paths with the top and bottom fibers changing from tension to compression and from compression to tension several times along the longitudinal direction in close accordance to what occurs as a result of wrinkling.

In this sense, Figure 3.13 can be seen as a consequence of the phenomenological interpretation of the two different results of Figure 3.12.

3.4.3 Sensitivity to material parameters

Finite element modeling of FRF using Strategy 2 allowed the authors to perform sensitivity analysis on the risk of wrinkling with respect to materials parameters such as the yield stress and the strain hardening coefficient.

In the case of the yield stress σ_Y, one-dimensional analytical models based on the plastic instability of plates will straightforwardly predict an increase of the compression forces applied in flanges and, therefore, an increase of the risk of wrinkling, when σ_Y increases. The evolution of effective strain vs. stress triaxiality in transverse and longitudinal cross sections of the profiles that is illustrated in Figure 3.14 corroborates the aforementioned one-dimensional predictions because small values $\sigma_Y = 150\,\mathrm{MPa}$

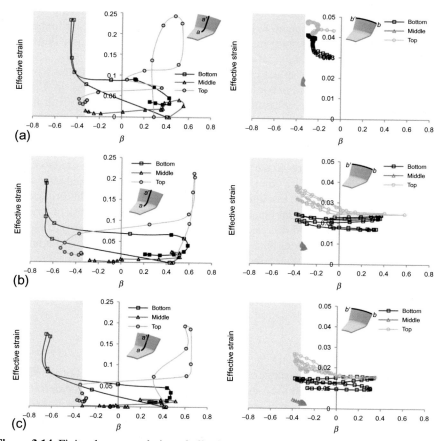

Figure 3.14 Finite element evolution of effective strain vs. stress triaxiality for the transverse (left) and longitudinal (right) directions of variable cross section profiles made from materials with different values of the yield stress. (a) $\sigma_Y = 150\,\mathrm{MPa}$, (b) $\sigma_Y = 400\,\mathrm{MPa}$, and (c) $\sigma_Y = 650\,\mathrm{MPa}$.

Table 3.4   Mechanical properties and Ludwik–Hollomon's parameters that were used in Strategy 2

Material number	Modulus of elasticity, E (GPa)	Yield strength, σ_Y (MPa)	Constant, K (MPa)	Strain hardening, n
1	200	150	632.53	0.2
2	200	400	1386.29	0.2
3	200	650	2044.26	0.2
4	200	400	1016.03	0.15
5	200	400	1891.48	0.25

will not give rise to wrinkling. However, such predictions are not able to explain the reason why critical values of stress triaxiality $\beta < -1/3$ for the development of wrinkling are more significant for $\sigma_Y = 400\,\text{MPa}$ than for $\sigma_Y = 650\,\text{MPa}$ (refer to Figure 3.14b and c). The explanation is based on the fact that the increase in the mean normal stress σ_m is smaller than the increase in the effective stress $\bar{\sigma}$ and will give rise to a decrease in the absolute value of stress triaxiality $\beta = \sigma_m / \bar{\sigma}$.

In order to investigate the effect of strain hardening exponent n on the wrinkling tendency, three different strain hardening exponents n were applied in Ludwik–Hollomon's equation, which by assuming a certain yield stress σ_Y will result in three different strength coefficients K (Table 3.4, Materials 2, 4, and 5). In these circumstances, the increase of n will give rise to an increase of K. Wang, Kinzel, and Altan (1994) proposed a wrinkling criterion for an elastic isotropic and plastic anisotropic shell with compound curvatures in the noncontact region of a sheet subjected to internal forming stresses. The wrinkling criterion for the shrink flange is obtained after simplification of the general criterion to a uniaxial stress state. According to this criterion, the critical strain at the onset of wrinkling decreases, or in other words the risk of wrinkling increases, by increasing K and decreasing n. The evolution of effective strain vs. stress triaxiality in transverse and longitudinal cross sections of the profiles in Figure 3.15 shows wrinkling to develop in all three cases and the risk is intensified by increasing n from 0.15 to 0.2. The decrease in the severity of wrinkling when n increases from 0.2 to 0.25 may be explained by the general influence of K over the tendency to wrinkling.

3.5   Conclusions

Two finite element modeling strategies combined with experimentation in variable cross section profiles made of St12 steel sheets allowed investigating the deformation mechanics of FRF. Combination of strains and stresses into loading paths plotted in the space of effective strain vs. stress triaxiality along transverse and longitudinal directions proved effective to explain material flow and failure by wrinkling. Wrinkling occurs when a significant cross sectional area of the flanges experience stress triaxiality levels below that of homogeneous compression $\beta = \sigma_m / \bar{\sigma} < -1/3$ and will result in "wave-type" loading paths with the outer and inner fibers changing from tension to

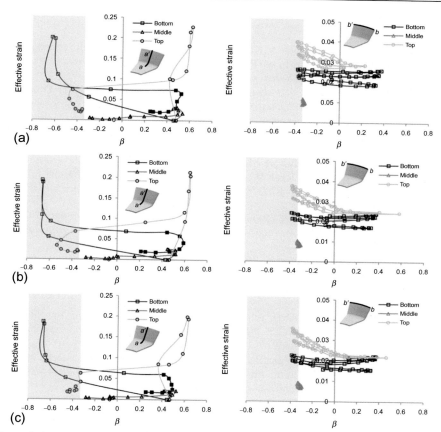

Figure 3.15 Finite element evolution of effective strain vs. stress triaxiality for the transverse (left) and longitudinal (right) directions of variable cross section profiles made from materials with different strain hardening coefficients. (a) $n = 0.15$, (b) $n = 0.2$, and (c) $n = 0.25$.

compression and from compression to tension several times along the longitudinal direction. Material parameters such as the yield stress σ_Y, the strength coefficient K, and the strain hardening coefficient n are also found to influence the risk of wrinkling with smaller values being more adequate to avoid failure.

Acknowledgments

The authors (M.M. Kasaei, H. Moslemi Naeini, and B. Abbaszadeh) would like to acknowledge the financial support provided by the Iranian Presidential Center for Innovation and Technology Cooperation (CITC) and would also like to thank Zolghar Industrial Factory for helping in the manufacturing of the FRF setup.

The work of M.Sc. Mehran Mohammadi during design and manufacturing of the FRF setup is greatly acknowledged. P.A.F. Martins and M.B. Silva would like to acknowledge the support provided by the Portuguese Foundation for Science and Technology.

References

Abee, A., Berner, S., & Sedlmaier, A. (2008). Accuracy improvement of roll formed profiles with variable cross sections. In *ICTP 2008—The 9th international conference on technology of plasticity, Gyeungju, Korea.*

Alvarez, W. (2006). *Roll form tool design fundamentals.* New York, NY: Industrial Press.

Azizi Tafti, R. (2014). *Theoretical, numerical, and experimental investigation of edge wrinkling defect in cold roll forming of symmetric channel sections.* PhD dissertation, Tehran, Iran: Tarbiat Moadares University.

Gao, J. F., Li, Q., & Zhao, W. (2013). Thermal stress analysis for local heating variable cross-section roll forming. *Advanced Materials Research, 683,* 599–603.

Groche, P., von Breitenbach, G., Jckel, M., & Zettler, A. (2003). New tooling concepts for future roll forming applications. In *ICIT 2003—The 4th international conference on industrial tools, Maribor, Slovenia.*

Groche, P., Zettler, A., & Berner, S. (2006). Analytic one-step-model for the design of flexible roll formed parts. In *Esaform 2006—The 9th international conference on material forming, Glasgow, UK.*

Gülçeken, E., Abeé, A., Sedlmaier, A., & Livatyali, H. (2007). Finite element simulation of flexible roll forming: A case study on variable width U channel. In *The 4th international conference and exhibition on design and production of machines and dies/molds, Cesme, Turkey.*

Hennig, R., Sedlmaier, A., & Abée, A. (2011). Finite element analysis of 3D-profiles with changing cross sections. In *ICTP 2011—The 10th international conference on technology of plasticity, Aachen, Germany.*

Kasaei, M. M., Naeini, H. M., Abbaszadeh, B., Mohammadi, M., Ghodsi, M., Kiuchi, M., et al. (2014). Flange wrinkling in the flexible roll forming process. In *ICTP 2014—The 11th international conference on technology of plasticity, Nagoya, Japan.*

Larrañaga, J., & Galdos, L. (2009). Geometrical accuracy improvement of flexible roll formed profiles by means of local heating. In *RollFORM09—1st international congress on roll forming, Bilbao, Spain.*

Larrañaga, J., Galdos, L., García, C., Ortubay, R., & Arrizabalaga, G. (2008). Flexible roll forming process reliability and optimization methods. In *ICTP 2008—The 9th international conference on technology of plasticity, Gyeungju, Korea.*

Lindgren, M. (2009). 3D roll-forming of hat-profile with variable depth and width. In *RollFORM09—1st international congress on roll forming, Bilbao, Spain.*

Ona, H. (2005). Study on development of intelligent roll forming machine. In *ICTP 2005—The 8th international conference on technology of plasticity, Verona, Italy.*

Ona, H., Sho, R., Nagamachi, T., & Hoshi, K. (2010). Development of flexible cold roll forming machine controlled by PLC. In *The 13th international conference on metal forming, Toyohashi, Japan.*

Park, J. C., Yang, D. Y., Cha, M., Kim, D., & Nam, J. B. (2014). Investigation of a new incremental counter forming in flexible roll forming to manufacture accurate profiles with variable cross-sections. *International Journal of Machine Tools and Manufacture, 86,* 68–80.

Tekkaya, A. E., & Martins, P. A. F. (2009). Accuracy, reliability and validity of finite element analysis in metal forming: A users perspective. *Engineering Computations, 26,* 1026–1055.

Vogler, F., Duschka, A., & Groche, P. (2010). Part accuracy of hollow profiles manufactured through flexible roll forming. *Steel Research International, 81*(9), 54–57.

Wang, C. T., Kinzel, G., & Altan, T. (1994). Wrinkling criterion for an anisotropic shell with compound curvatures in sheet forming. *International Journal of Mechanical Sciences, 36*(10), 945–960.

Yan, Y., & Li, Q. (2011). FEM modeling and mechanics analysis of flexible roll forming. *Applied Mechanics and Materials, 44*, 132–137.

Yan, Y., Wang, H., Li, Q., Qian, B., & Mpofu, K. (2014). Simulation and experimental verification of flexible roll forming of steel sheets. *International Journal of Advanced Manufacturing Technology, 72*, 209–220.

Zhao, W., Yan, Y., Wang, H. B., & Gao, J. F. (2013). Finite element analysis and fracture forecast of U channel flexible roll forming. *Advanced Materials Research, 683*, 604–607.

Research issues in the laser sheet bending process

R. Kant, S.N. Joshi, U.S. Dixit
Indian Institute of Technology Guwahati, Guwahati, India

4.1 Introduction

Due to the ability of precise focusing and ease in controlling the heat source, lasers are widely used in the manufacture of critical components of consumer electronics, aerospace systems and vehicles, biomedical instrumentation, automobiles, and in shipbuilding applications. Lasers can process a wide range of materials *viz.* metals, nonmetals, composites, and ceramics in a variety of ways such as heat treating, cutting, forming, engraving, marking, joining, cladding, sintering, scribing, drilling, etc. Sheet bending is an important forming process which has applications in manufacturing of macro- as well as micro-sized parts. The conventional method of sheet bending employs mechanical punches and dies to produce the required bend angle in the work sheet. The process is economical and generally used for mass production of components. However, the capital cost of the required equipment is generally very high. Moreover, a lot of time is needed in alteration or adjustment of tooling. Because of this, mechanical bending is not suitable for prototyping or low volume production. The springback effect and difficulty in forming complex shapes further beset its application in precision manufacturing of sheet metal components. Application of lasers for bending the metal sheets helps to overcome some of these limitations. The details of the laser bending process mechanism, process parameters, and research issues such as numerical modeling of the laser bending process, development of coatings for enhancing the coefficient of absorption, application of moving mechanical load in addition to laser beam energy, inverse analysis, and optimization are provided in the following sections.

4.2 Laser bending process

Laser bending is similar to flame bending, which is used for the adjustment of welded construction in ship building industries. The forming of metal plates with the application of flame heating was initiated at the start of the nineteenth century to shape the external metal plates of a ship's hull (Ojeda & Grej, 2009). This process is man–hour intensive and dependent on the skill of personnel. It is difficult to control and focus the flame on a small area. The problem can be solved by applying a laser beam instead of a gas flame to deform the metal sheets. The application of a laser as a heat source for sheet bending was first attempted by Kitamura in the early 1980s (Kitamura, 1983).

Materials Forming and Machining. http://dx.doi.org/10.1016/B978-0-85709-483-4.00004-1

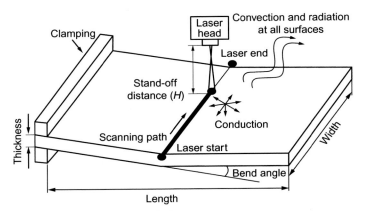

Figure 4.1 Schematic of laser beam irradiation.

The straight line irradiation process to produce a bend angle in a flat sheet metal piece is shown in Figure 4.1. The sheet is clamped at one side on a computer numerical control (CNC) machine and the other side is free or sometimes preloaded in case of laser-assisted bending. The heating on the material surface takes place due to scanning of the laser beam along a predefined path. Either the CNC table or the laser head or both of them simultaneously move to generate a specified straight or curvilinear scanning path. The laser beam irradiation causes rapid localized heating which is followed by cooling as the laser energy is moved onto an adjacent area. During heating, the expansion in the irradiated zone occurs and is resisted by the surrounding material. It results in thermal compressive stresses in the irradiated zone and tensile stresses in the surrounding surface of the work sheet. During natural cooling, the irradiated material undergoes shrinkage, leading to the development of bending or a change of shape of the work sheet at the irradiated region (Shi, Yao, Shen, & Hu, 2006). Various laser forming mechanisms are presented in the following subsections. These mechanisms can activate separately or in combination.

4.2.1 Temperature gradient mechanism

The temperature gradient mechanism (TGM) is mainly used to produce precise small bend angles in thick sheets. It basically depends upon the steep temperature gradient generated across the sheet metal thickness due to an intense laser beam applied on the work sheet.

TGM occurs when there is a steep temperature gradient along the worksheet thickness direction. In this mechanism, beam diameter is typically of the order of worksheet thickness. The feed rate is large enough to maintain a steep temperature gradient. The feed rate should be kept higher for the materials with higher thermal conductivity (Li & Yao, 2001). The bend angles between 0.1° and 3° can be achieved in a single laser pass (Lawrence, Schmidt, & Li, 2001). A typical laser bent specimen of magnesium alloy with TGM is shown in Figure 4.2.

Figure 4.2 Laser bent specimen with TGM (500 W power, 3000 mm/min scan speed, 3.87 mm laser spot diameter, $60 \times 40 \times 1.9$ mm^3 sheet).

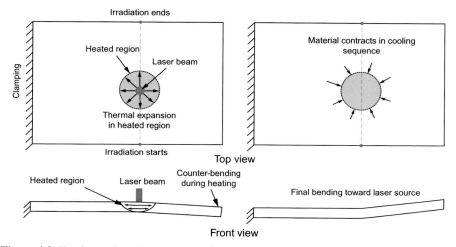

Figure 4.3 Heating and cooling sequences in temperature gradient mechanism (TGM).

Figure 4.3 shows the sequences involved in TGM. TGM occurs in two sequences, heating and cooling of the work sheet. During heating, the thermal expansion occurs in the heating region which causes bending of the work sheet away from the laser source. This is called counter bending. The thermal expansion is resisted by the surrounding cooler material. It generates compressive stresses in the heated region and tensile stresses in the surrounding cooler region. The yield stress and Young's modulus drastically reduce at high temperature. When induced thermal stresses exceed temperature dependent flow stress, plastic deformation occurs. The plastic deformation is compressive in the heated region, and therefore the top irradiated surface gets compressively deformed. The plastic deformation at the bottom surface is negligible due to the lower temperature. The material contracts during cooling. The compressive plastic deformation causes local shortening at the top irradiated region. Thus, the work sheet finally bends toward the laser source as shown in Figure 4.3. The bending is a result of the difference between plastic deformation at top and bottom surfaces (Kant & Joshi, 2012a).

Aghyad (2014) performed numerical investigation on laser bending of D36 shipbuilding steel sheet with the help of the finite element method (FEM). The relation between temperature and bend angle histories was studied to explore the TGM. Figure 4.4 shows a large temperature difference between the top and bottom surfaces of the sheet. The large gradient along the thickness direction generates the plastic

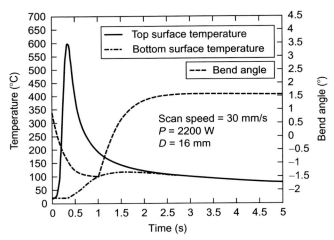

Figure 4.4 Comparison between temperature and bend angle histories (Aghyad, 2014).

strains, which produces the bending of the metal sheet. It can also be seen that the counter-bending of about 1.5° occurs at time 1 s. It increases until the bottom surface attains the peak temperature. As the temperature at the bottom surface decreases, the bend angle starts increasing and then it flattens at its maximum value. At this moment, the temperatures at the top and bottom surfaces become equal.

4.2.2 Buckling mechanism

The buckling mechanism (BM) is a mechanism used for bending of thin metal sheets. In BM, the temperature at the top and bottom surfaces is almost equal, thus temperature gradient is negligible compared to that in TGM. The BM occurs when thin sheets of high thermal conductivity such as aluminum are irradiated with a slow speed laser beam of a diameter larger than the sheet thickness (approximately 10 times). Figure 4.5 shows the various process sequences involved in the BM. Unlike in TGM, here the counter-bending does not occur during the heating sequence. Instead the bending direction depends on the precurvature of the sheet, internal stresses, and external or gravitational forces (Shi et al., 2006).

The negligible temperature gradient in BM causes uniform thermal strain throughout the thickness as shown in Figure 4.5a. The free expansion of the heated region is restricted by the surrounding cooler material. It results in compressive stresses in the heated region. The buckling stiffness of the sheet is low due to less thickness and lateral expansion caused by the temperature field is large. It creates buckle in the heated region as shown in Figure 4.5b. The tendency to buckle is enhanced when the sheet is thin and the coefficient of thermal expansion and temperature dependent flow stress are high. Once buckling is initiated, it extends along with the laser irradiation. When thermal stresses exceed temperature dependent flow stress, the plastic deformation occurs in the buckle and the work sheet bends as shown in Figure 4.5c. In this way, the bend angles are generated in the BM-dominated laser bending process. The direction

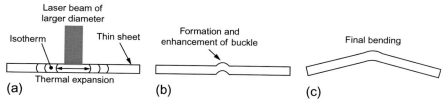

Figure 4.5 Process sequences of buckling mechanism (BM): (a) initial heating, (b) growth of buckle, and (c) development of bend angle.

of the buckling is predetermined by the existing buckle and the remaining part buckles in the same direction. The bend angle of the order of 15° in one pass can be produced by using BM. A typical laser bent specimen of aluminum bent with the laser bending BM is shown in Figure 4.6.

Li and Yao (2001) postulated an irradiation scheme which starts from a location near the middle of the work sheet instead of, as normally, from an edge of the work sheet. The proposed irradiation scheme is shown in Figure 4.7. First rightward and then leftward scanning was done from near the middle of the scanning path. The proposed scheme was able to provide convex bending in the BM-dominated process mechanism. The shift of the starting point to the middle added mechanical constraints to sustain the initial convex deformation. Using the proposed irradiation scheme, the convex bending can be achieved without application of prebending or external mechanical constraints.

The critical condition to know whether the process is dominated by the thermal gradient mechanism or by the BM can be given as (Shi et al., 2006):

$$\frac{Pd^{1/2}}{h^2V^{1/2}} > \frac{\eta \pi^{7/2} k^{1/2} \rho^{1/2} c^{1/2}}{41.52(1+\mu)A\alpha_{th}} \tag{4.1}$$

where P, d, h, V, η, k, ρ, c, μ, A, and α_{th} are laser power, beam diameter, sheet thickness, scanning velocity, correction factor, thermal conductivity, density, specific heat, Poisson's ratio, absorption coefficient, and coefficient of thermal expansion, respectively. The parameters satisfying the condition in Equation (4.1) provide BM-dominated

Figure 4.6 Laser bent specimen with BM (200 W power, 300 mm/min scan speed, 6 mm laser spot diameter, $80 \times 50 \times 0.5$ mm³ sheet).

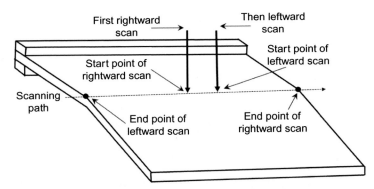

Figure 4.7 Irradiation scheme for a certain laser bending direction in BM.
Redrawn from Li and Yao (2001), with permission. Copyright 2001 Elsevier.

process conditions. The process condition for the dominating mechanism can also be assessed by a Fourier number which is given as (Shi, Liu, Yao, & Shen, 2008):

$$Fo = \frac{\alpha_d d}{h^2 V} \tag{4.2}$$

where α_d, h, d, and V are thermal diffusivity, sheet thickness, beam diameter, and scanning velocity, respectively. The smaller value of the Fourier number corresponds to a TGM-dominated laser bending while BM plays a dominant role for a high Fourier number.

4.2.3 Upsetting mechanism

The upsetting mechanism (UM) is also called a shortening mechanism. The UM occurs if the laser beam diameter is of the order of, or less than, the sheet thickness, the processing speed is low, and the thermal conductivity is high. The process condition of the UM is almost the same as the BM and the temperature gradient in worksheet thickness is negligible. The only difference is the beam diameter which is less than the worksheet thickness. It makes worksheet geometry stiff enough to prevent buckling. The UM is illustrated in Figure 4.8.

In the UM, temperature gradient is negligible; therefore, the thermal expansion due to laser heating is almost equal at the top and bottom surfaces. This expansion is

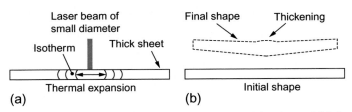

Figure 4.8 Process steps of the upsetting mechanism (UM): (a) heating and (b) final shape.

resisted by surrounding cooler bulk material which generates compressive stresses in the heated region. The geometry of the work sheet is stiff, which prevents the sheet from buckling. When induced thermal stresses exceed temperature dependent flow stress, the plastic deformation occurs. The plastic deformation is almost equal at the top and bottom surfaces and is compressive in nature. During cooling, it results in local shortening and thickening of the work sheet (Li & Yao, 2001; Shi et al., 2006). With a proper heating strategy, the mechanism can also be used to form a plane sheet into a specially formed part. The mechanism can also be used for the shortening of small frames, pipe bending of various kinds of cross-sections, and aligning in micro-parts operations.

4.3   Advantages, disadvantages, and applications of laser bending

The laser bending has the following advantages in comparison with that of the conventional bending operation:

1. The process is flexible and easy to control.
2. External force, tools, dies, and presses are not required.
3. The process is faster due to elimination of lead time associated with mechanical bending because special hard tools are not required.
4. A precise and small bend angle can easily be obtained.
5. Forming at inaccessible areas is possible as the laser transmission is easy into the inaccessible areas.
6. At elevated temperatures, material's formability increases; therefore, the difficult-to-form materials such as magnesium as well as brittle, hard, and thick material can be processed.
7. The heat affected zone is small as the laser beam has a narrow, concentrated, and controlled area of irradiation. Thus, worksheet degradation due to heating is less.
8. Complex shapes can be generated with selected irradiation strategies.

In spite of various advantages, laser bending has some limitations. These are as follows:

1. In case of mass production of large sized sheet metal components, the laser bending process is slow in comparison with the traditional punch and die method.
2. The process is not suitable for materials having high reflectivity.
3. In case of improper selection of laser parameters, melting may occur. This degrades the work surface.
4. The material and microstructural properties of an irradiated region may deteriorate.
5. Bend angle per irradiation is small and, therefore, to get a higher bend angle, multiple irradiations are required. This leads to an increase in energy consumption and loss of efficiency.

Laser bending has applications in several fields of industrial manufacturing which include automotive, aerospace, shipbuilding, medical, microelectronics, and material processing. Laser bending is used to manufacture small and precise bend angles in very small parts which are used for microelectro mechanical systems, chemical, and sensor industries. Laser bending is used as an accurate and cost effective process to adjust or align mating parts in welded constructions, the ship building industry, and in straightening of car body parts. It is used in the generation of complex smooth surfaces, rapid prototyping, and shape correction of bodies made of sheet metal. It is also used for bending of brittle materials.

4.4   Process and performance parameters

During laser bending, the desired mechanism can be achieved by controlling various process parameters. The process parameters can be categorized into three groups—energy parameters, material properties, and worksheet geometry. These are listed in Table 4.1.

4.4.1   Laser power

Laser power controls the amount of energy absorbed into the work sheet. In general, the bend angle increases with an increase in laser power, attains a peak, and then decreases with further increase in the laser power. The increase in bend angle with laser power is due to more heat absorption which causes a higher peak temperature and hence more plastic deformation at the scanned surface (Kant, Joshi, & Dixit, 2013b). After attaining peak, the bend angle decreases with an increase in laser power which is mainly due to two reasons. First, the melting occurs in the irradiated region at higher power and the applied heat energy gets consumed in the phase transformation instead of worksheet bending (Lawrence et al., 2001). Second, at high power, the peak temperature at the bottom surface is also high. This reduces the difference between plastic deformation at the top and bottom surface which leads to a decrease in bend angle at higher power (Kant & Joshi, 2012a). The laser bending below a certain laser power is not possible due to reversible elastic effect or the threshold energy (Chen, Wu, & Li, 2004a).

4.4.2   Scanning speed or feed rate

The scanning speed or feed rate is an important parameter which controls the contact time between the laser beam and the worksheet surface. The bend angle nonlinearly varies with scanning velocity. The bend angle first increases, attains a peak, and then

Table 4.1   **Parameters affecting the laser bending process**

Energy parameters	Material properties	Worksheet geometry
Laser power	Coefficient of thermal expansion	Thickness
Scanning velocity	Thermal conduction	Width
Beam diameter	Specific heat	Length
Number of scans	Elastic modulus	
Cooling condition	Absorption coefficient	
Scanning path	Poisson's ratio	

decreases with increase in scanning speed. This happens because the significant temperature gradient cannot be maintained at a very low scanning velocity. At moderate scanning velocity, the temperature gradient is more; thus the bend angle is high. However, if the speed increases further the bend angle reduces. This is due to low energy input per unit length.

4.4.3 Beam diameter

The laser beam diameter is the measured diameter of the irradiation spot at the worksheet surface. It can be changed by varying the stand-off distance between the laser head and the worksheet. The beam diameter increases with an increase in stand-off distance. The heat flux density and temperature gradient both decrease with an increase in laser beam diameter. Decrease in heat flux density and temperature gradient both result in production of lower bend angles at higher beam diameter.

4.4.4 Absorption coefficient

During the laser bending process, the absorption behavior has a complex interaction with various process parameters. The absorptivity increases when a laser beam of shorter wavelength (e.g., Nd:YAG) is used. The absorptivity also depends on the worksheet surface condition. It increases with surface roughness due to multiple reflections of the beam. The metals have the property of high reflectivity but the presence of rust and oxide layers increases the absorptivity. The absorptivity can be increased by applying a surface coating. The graphite coating is preferred due to high absorptivity (60–80%) and simple coating technique. Dutta, Kalita, and Dixit (2013) found that the bend angle is significantly high when a simple black enamel paint coating is applied. They also found that coating damages due to laser irradiation and the bend angle was higher when coating was applied after each irradiation. The effect of coating is shown in Figure 4.9.

Singh (2013) observed that lime coating is very effective in the laser bending process. He also applied graphite grease coating. It was observed that the coefficient of absorption increased by a factor of two in the case of the graphite grease coating, while it increased up to a factor of fifteen in the case of lime coating. Lime coating is inexpensive and environmentally friendly.

4.4.5 Number of scans

The laser bending with more than one irradiation is called as multipass laser bending process. The bend angle increases with the number of irradiations. The effect of irradiations is not uniform and, generally, the increment in bend angle after each irradiation decreases. This is because in each successive irradiation, the temperature gradient decreases as the work sheet is already preheated due to the previous irradiation (Kant & Joshi, 2013b). The change in worksheet thickness, strain hardening, and change in beam geometry due to worksheet bending are also important parameters which reduce the change in bend angle after each irradiation (Edwardson, Abed, Bartkowiak,

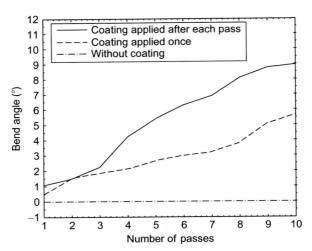

Figure 4.9 Variation of bend angle with number of passes for low power–high speed (500 W power, 1300 mm/min scan speed) (Dutta et al., 2013).

Dearden, & Watkins, 2006). The continuous irradiations may also lead to melting of the work sheet. Therefore, proper selection of process parameters and cooling conditions is required for an efficient and quality multipass laser bending process. The damage of surface coating is also one of the important challenges in laser bending with multiple irradiations.

4.4.6 Cooling condition

The cooling condition is a parameter that influences the laser bending process during multipass operation. In a multipass laser bending process, significant waiting time is necessary to cool down the work sheet so that a steep temperature gradient can be achieved in the next irradiation. The forced cooling after each irradiation can reduce this waiting time significantly. It helps in avoiding surface melting after a few irradiations and maintaining higher temperature gradient in each successive laser scan. It leads to the increase of change in bend angle after each successive pass (Kant & Joshi, 2013c). Aghyad (2014) showed by FEM that the cooling rate has an insignificant effect on the final bend angle. Hence, it is a good idea to cool the sheet after each pass.

Shen, Hu, and Yao (2011) investigated the effect of forced water cooling at top, bottom, and at both the surfaces simultaneously during single-pass laser bending of steel plate. They reported a significant improvement in bend angle when the cooling was applied at top and bottom surfaces simultaneously in comparison to other cases.

4.4.7 Effect of material properties

Laser bending is a thermomechanical process and, therefore, both thermal and mechanical properties of sheet affect the process. The important thermal properties are thermal conductivity (k), heat capacity (c_p), density (ρ), and the coefficient of thermal expansion (α). The material parameters mainly relate to a thermal effect index (R), in other words, the ratio of the coefficient of thermal expansion to the volumetric heat capacity (expressed by the product of the specific heat and the density) which can be written as $R = \alpha/\rho c_p$. The bend angle increases with index R in a piecewise linear fashion (Shichun & Jinsong, 2001). Thermal conductivity determines magnitude of temperature gradient and thus decides the dominating mechanism. The bend angle decreases with an increase in thermal conductivity. This is because the peak temperature and temperature gradient decreases due to quick heat dissipation in high conductivity materials.

The effects of mechanical properties can be described by ratio of flow stress to elastic modulus. This ratio is equal to the elastic strain at yield strength. Thus, the irreversible elastic strain is small for low strength materials. This results in an effective conversion of thermal expansion into a plastic strain. On the contrary, the high strength material (e.g., titanium alloy) shows a high ratio of the flow stress to the elastic modulus. Thus, the reversible elastic strains are more in the high strength material which results in low bend angles compared to that in low strength material (Duflou, Callebaut, Verbert, & Baerdemaeker, 2007).

4.4.8 Effect of worksheet geometry

Work sheet thickness, length, and width are to be considered as worksheet geometry parameters during the laser bending process. The sheet thickness is the most important parameter as it directly controls the temperature gradient and material constraints. The bend angle is approximately inversely proportional to the square of the sheet thickness (Geiger, Vollertsen, & Deinzer, 1993). The counter-bending is less for a wider sheet because of rigid-end effect. The bend angle increases with an increase in the sheet width (Shichun & Jinsong, 2001). This is because an expansion along the width is possible for a work sheet of lower width which decreases stress in the forming zone. It effectively decreases the process efficiency. The sheet length does not have much influence on bend angle (Chen et al., 2004a).

4.5 Effect on microstructural and mechanical properties

The study of effect of laser parameters on microstructural and mechanical properties is an important aspect for manufacturing of defect free and good quality products. Very few experimental studies are carried out to investigate microstructural and mechanical properties of a laser bent specimen.

4.5.1 Microstructural studies

The laser bending process involves high temperature and localized distortions near the irradiated region which may cause a change in metallurgical properties. The effects of laser irradiation may be different for different materials. Hu, Wang, Labudovic, and Kovacevic (2001) analyzed the surface of the laser irradiated stainless steel plate and observed that it does not have any harmful effects on the microstructure of the plate and does not produce any cracks or porosity in the plate. However, sometimes a few scattered microcracks can be generated on an irradiated surface which are associated with a high cooling rate (Yilbas, Arif, & Abdul Aleem, 2012).

Chen et al. (2004a) analyzed the microstructures of Ti–6Al–4V alloy and did not find any obvious difference when compared with the microstructure of the original state. It was also found that the scanning at high laser energy density separated out the α-phase from β-grain boundaries, and β-grain size had increased. Fan, Yang, Cheng, Egland, and Yao (2005) studied phase transformations in the heat affected zone of AISI 1010 stainless steel. They found that grains were distinctly refined due to phase transformation and significant recrystallization occurred in the heat affected zone. After irradiation, a substantial amount of martensite was formed due to a high cooling rate.

Singh, Joshi, Ray, and Dixit (2013) studied the microstructure of a multipass laser bent mild steel sheet. They found that the average grain size gradually decreases from the bottom to the top surface of the irradiated region. The microstructures at the top and bottom surfaces are shown in Figure 4.10a and b. The reformation from coarse grain to fine grain was more at the top surface than the bottom surface. It was due to a higher temperature and higher strain hardening at the top surface.

4.5.2 *Mechanical properties*

The laser bending is a high temperature phenomenon. It involves strain hardening, compressive and tensile deformations, and generation of residual stresses in the irradiated region. These factors affect the mechanical properties like tensile strength, fatigue strength, hardness, and ductility of the material.

McGrath and Hughes (2007) found that the fatigue life of worksheet materials was substantially enhanced after the laser beam irradiations. The endurance limit of the laser-formed specimen also increased. It was due to the laser-hardening mechanism and compressive residual stresses induced due to laser beam irradiations. Shen and Yao (2009) observed improvements in the yield strength and tensile strength of

Figure 4.10 Microstructure after 20 laser passes (400 W power, 300 mm/min scan speed) for (a) top surface and (b) bottom surface (Singh et al., 2013).

a laser-formed low carbon steel specimen, while the percentage of elongation was reduced. Cheng and Yao (2001) found that the cooling in multipass laser bending of stainless steel moderately decreased the material ductility because the repeated work hardening was offset by repeated softening. The hardening occurred due to plastic deformation and the softening occurred because of recovery and recrystallization accompanying each laser scan.

Fan et al. (2005) worked on AISI 1010 stainless steel and found that microhardness decreases along the thickness of the laser bent specimen. Hardness was found to be influenced by both phase constitution and work hardening. Due to the high cooling rate, the substantial amount of martensite was formed at the top surface which has high hardness. Also, the top surface experienced a higher peak temperature and larger plastic deformation, and thus, has higher hardness due to work hardening.

Singh et al. (2013) studied the microhardness of a laser bent mild steel specimen and compared it with a mechanically bent specimen. In a laser bent specimen, hardness increased at the top surface while it decreased at the bottom surface. This was because the top surface has a higher temperature and undergoes hardening due to air cooling while the bottom surface has a low temperature which is not sufficient for hardening. However, the hardness of the bottom surface was also higher than the parent material due to the strain hardening effect. The hardness increased with laser power due to the higher peak temperature. At low power, an increase in scanning speed reduced microhardness and this trend reversed at higher laser powers. It was because the attainable microhardness is a function of time and temperature. The scan speed controls the heating time and to some extent the rate of cooling. At a low power, higher heating time favors the hardening while at higher powers, lower heating time favors the hardening. Singh et al. (2013) also carried out a 3-point flexure test on the bent sheets. The flexural stiffness of the laser bent sheet was found to be more as compared to a mechanically bent sheet as shown in Figure 4.11. It is seen from the figure that

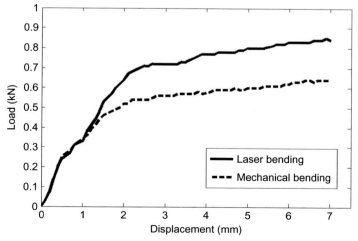

Figure 4.11 Comparison of the strength of a laser bent specimen after 10 laser passes (400 W power, 300 mm/min scan speed) and a mechanically bent specimen (Singh et al., 2013).

initially the stiffness of both the sheets is the same, but later the load required to cause deflection increases for the laser bent sheet. This indicates that Young's modulus of elasticity is the same for both sheets, but the elasto-plastic stiffness of the laser bent sheet is more than that of the mechanically bent sheet.

4.6 Curvilinear laser bending

The curvilinear laser bending can be used to create complex profiles. Various complex shapes can be generated with selected curvilinear irradiations. The deformation behavior of curvilinear laser bending is different from the straight line laser bending process. Chen, Wu, and Li (2004b) studied the deformation behavior of laser curve bending of a sheet and found that the deformation occurs only on one side of the scanning path along which the rigid constraint is relatively lower. Zhang, Guo, Shan, and Ji (2007) found that the peak temperature at the scanning surface and warped curvature increases with an increase in the scanning path curvature.

During their numerical and experimental studies, Kant and Joshi (2014) found that the bending occurred outside the scanning path curvature as shown in Figures 4.12 and 4.13. The bending was offset near the edges. This behavior is quite different from the bending due to straight line irradiation. In straight line irradiation, bending occurs along the irradiation line.

In curvilinear irradiation, the bending offset increased with an increase in beam diameter and laser power. It was also noted that the bend angle increases with an increase in the scanning path curvature. This is due to absorption of more energy along the longer scanning path with the higher scanning path curvature.

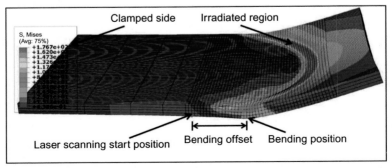

Figure 4.12 Bending behavior in the curvilinear laser bending process (Kant & Joshi, 2014). Published by Sage Publications, Copyright with authors.

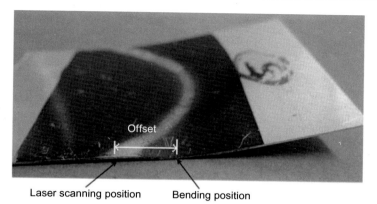

Figure 4.13 Laser bent aluminum specimen with curvilinear irradiations (250 W power, 300 mm/min scan speed) (Kant & Joshi, 2012b).

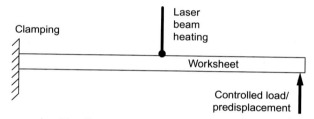

Figure 4.14 Laser-assisted bending.

4.7 Laser-assisted bending

Laser-assisted bending is a variant of laser bending. In this process, the formability of the work sheet is enhanced by heating with the laser beam, while the plastic deformation is carried out by a conventional mechanical tool. On the other hand, in a pure laser bending process, the deformation occurs only due to the thermal stresses induced by laser irradiations. Figure 4.14 shows a schematic of an arrangement of laser-assisted bending with controlled load or predisplacement.

A single-pass laser bending process has a limitation: the bend angle per pass is small (of the order of 1°). Also, the bending direction is not certain in the BM-dominated laser bending process. Therefore, single-pass, noncontact laser bending may not be suitable for forming of larger sized (of the order of 1 m) work sheets at mass scale. One possible solution can be the use of external mechanical force along with laser irradiations which may be called laser bending with a moving mechanical load. It bends the work sheet by the thermal as well as mechanical stresses. Laser heating enhances the

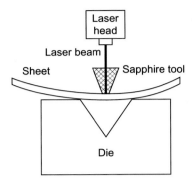

Figure 4.15 Schematic of laser-assisted bending using a sapphire tool.

formability in terms of reduction in yield strength and hardening coefficient. However, a small amount of springback is present and the external experimental setup is required to apply the controlled mechanical forces.

The external force/load is an important parameter that increases the bend angle and the certainty of bending direction during the laser-assisted bending process. Sometimes, to transmit laser light to a worksheet surface, sapphire tools are used (Gillner et al., 2005). The sapphires can transmit lasers and possess excellent mechanical properties such as high hardness and compressive strength. The work sheet can directly be heated at the area where material flow is required. The sapphire tool is used to carry out the plastic deformation. Figure 4.15 shows the schematic of laser-assisted bending with a sapphire tool.

Yao, Shen, Shi, and Hu (2007) carried out numerical investigation on the effects of various preloads on the bend angle. Four different preloading cases were studied. These were pure compression, pure tension, pure bending toward the laser beam, and pure bending away from the laser beam. The results showed that pure compression and pure bending (toward the laser beam) increased the bend angle, while pure tension and pure bending (away from the laser beam) decreased the bend angle. Liu, Sun, Guan, and Ji (2010) found that the bending direction in the BM-dominated process can be controlled by changing the direction of the prestresses.

For the application of larger sheets, the laser-assisted bending with a moving mechanical load was studied by Kant and Joshi (2013a). Laser bending with the simultaneously moving load methodology was developed to produce large bend angles of the order of 10°. Figure 4.16 shows the schematic of the developed experimental setup.

The setup is rigidly fixed and does not move with the laser machine bed. The load is applied in terms of predisplacement using a height adjustment screw. The setup has a rolling contact with the sheet through a roller. The roller arrangement is provided to reduce the friction and wear during the application of a moving load. The work sheet is clamped over the laser machine bed and moves along with the bed. The laser source and load arrangement are stationary and the sheet moves with a predefined velocity. In this way, a synchronized moving laser beam and moving mechanical load is applied to the work sheet. A study on bending mechanism, bend angle, springback effect, edge effect, and edge displacement were carried out for bending of magnesium work sheets.

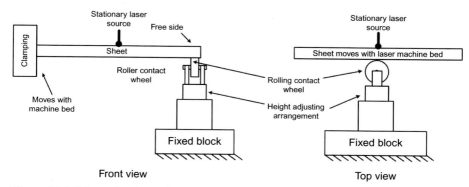

Figure 4.16 Schematic of experimental setup for laser-assisted bending with a moving mechanical load.

Magnesium has limited formability. The bend angle was found to be increased with an increase in mechanical load. The presence of springback and edge effect was also observed. During the experiments on laser bending of magnesium alloy M1A using CO_2 lasers, it was observed that with an increase in predisplacement from 5 to 10 mm, the bend angle gets doubled.

4.8   Edge effect and its control

In the laser bending process, the material boundaries are not uniform along the irradiation path. The material distribution around the laser beam is more uniform when the laser beam is at the middle of the work sheet and it is negligible at the free side when the laser beam is near the edges. It results in nonuniform temperature distribution and material constraint along the irradiation path. Therefore, the bend angle is not uniform along the laser irradiation and it varies from one end to the other of irradiation path. This is called an edge effect (Kant, Joshi, & Dixit, 2013a). The edge effect is generally undesirable, but in some cases it can be utilized to generate complex shapes and alignments. The main reasons for edge effect are:

1. Peak temperature is more near the edges in comparison with that in the middle of the irradiation path. This is because the material volume for heat conduction at the edges is less.
2. The ratio of surrounding cooler region to heated region is not uniform along the irradiation path. It is the maximum when the laser beam starts irradiation and decreases as it moves forward. Also the amount of heated region increases as the laser beam irradiates and the material looses stiffness at an elevated temperature.

The edge effect can be controlled by resolving the above issues. Shen, Hu, and Yao (2010) showed that scanning speed is an important parameter to control the edge effect and the combination of acceleration and deceleration scan scheme can minimize the edge effects. Hu, Xu, and Dang (2013) proposed two methodologies to reduce the edge effect; first is to maintain constant peak temperature along the irradiation path and second is to put external mechanical constraint in the form of clamping at both ends of the irradiation path. These methodologies significantly reduced the edge effect. Zahrani and Marasi (2013) showed that the following parameters, in order of significance, directly affect the edge effect: number of irradiations, worksheet thickness, scanning speed, and laser power. They found that edge effect decreases with an increase in number of irradiations, sheet thickness, and scanning speed and it decreases with a decrease in beam diameter.

4.9 Mathematical modeling of laser bending process

There have been a number of attempts to develop mathematical models of the laser bending process. These models make use of several simplifying assumptions. A brief review of mathematical modeling of laser bending process is provided in this section.

Vollertsen (1994) derived an expression for the bending angle for TGM. The analytical expression is given by

$$\alpha_b = \frac{3\alpha_{th}P\eta}{\rho c_p v h^2} \tag{4.3}$$

where α_b is the bending angle, α_{th} is the coefficient of thermal expansion of the work piece, P is the laser power, η is the absorption coefficient, ρ is the density, c_p is the specific heat capacity, v is the velocity, and h is the sheet thickness. The Vollertsen's model does not include the effect of yield stress and Young's modulus of elasticity of material. Yau, Chan, and Lee (1997) included this effect. In their model, the bending angle is given by

$$\alpha_b = \frac{21\alpha_{th}P\eta}{2\rho c_p v h^2} - \frac{36l\sigma_y}{hE} \tag{4.4}$$

where l is the half length of heated zone, E is the Young's modulus, and σ_y is the yield stress.

Kyrsanidi, Kermanidis, and Pantelakis (2000) considered nonuniform temperature distribution throughout the thickness of the plate and applied the concept of basic mechanics of materials. Their model, although computationally efficient, requires programming and includes iterative steps. Cheng et al. (2006) proposed an analytical model for plate with varying thickness. The bending angle of the plate at the location with thickness, $h(x)$, is

$$\alpha_b = b\left(1-v^2\right)\varepsilon_{max}\left(\frac{3f(x)\pi}{2h^2(x)} - \frac{4f^2(x)}{h^3(x)}\right) \tag{4.5}$$

where

$$b = c_1\sqrt{P/v} \tag{4.6}$$

and

$$f = c_2 P / (vh) \tag{4.7}$$

where v is scan speed, c_1 and c_2 are constants dependent on materials properties, $h(x)$ is the thickness of the sheet, and ν is the Poisson's ratio. In Equation (4.5), ε_{max} is the maximum plastic strain at the heated surface. It is given by

$$\varepsilon_{max} = \alpha_{th} \theta_{max} - \sigma_y / E \tag{4.8}$$

where θ_{max} is the maximum temperature increase.

The model of Shen, Yao, Shi, and Hu (2006) is based on the assumption that the plastic deformation is generated only during heating, while during cooling the plate undergoes only elastic deformation. According to this model, the bending angle is given by

$$\alpha_b = \left[4\beta_v + \frac{12k\sigma_y}{E} \frac{2a_p r}{h(h - a_p)} \right] \frac{ha_p}{(h - a_p)^2} \tag{4.9}$$

where β_v is the bending angle provided by Equation (4.3), r is the laser beam radius, k is the reduction coefficient to account for the variation of yield strength and Young's modulus of elasticity with temperature, and a_p is the characteristic length of the plastic zone. The constants k and a_p need to be evaluated empirically, which is a limitation of this model. The model is valid for TGM as well as BM.

Lambiase (2012) proposed an expression for the bending angle based on assumption of the elastic-bending theory without taking into account plastic deformation during the heating and cooling phases. The bending angle is given by

$$\alpha_b = \frac{3\eta P (h - h_1) \alpha_{th}}{\rho v c_p h (h^2 - 3hh_1 + 3h_1)} \tag{4.10}$$

where h is sheet thickness and h_1 is the thickness of the heated volume, which is invariably estimated empirically. Lambiase and Ilio (2013) developed a more rigorous analytical model to predict the deformation of thin sheets.

Several other analytical models have been proposed. Shi, Shen, Yao, and Hu (2007) provided a model for estimating the bend angle in an in-plane axis perpendicular to the scan direction. Vollertsen (1994) provided an expression for estimating the bend angle during BM mechanism. Kraus (1997) provided a closed-form expression for estimating the bend angle during the UM.

The accuracy of closed-form expressions to predict laser bending is not high. It typically ranges from 10% to 50%. Moreover, the closed-form expressions only limited information. To get better insight, several FEM methods have been proposed (Chen & Xu, 2001; Chen, Xu, Poon, & Tam, 1999; Hu et al., 2001; Kyrsanidi, Kermanidis, & Pantelakis, 1999). However, FEM simulation takes several hours (typically 10 h for one simulation) and is not suitable for online optimization and control (Kyrsanidi et al., 2000). Some finite difference models have been proposed, which also require long computational times. Aghyad (2014) proposed a finite element model that employs accelerated cooling to reduce the computational time. He employed this methodology for modeling of stationary pulsed laser bending as well as scanning laser bending of strips. The methodology is implemented in the FEM package ABAQUS. He also suggested stopping the computations when the top and bottom surface temperatures become equal. This cuts down a lot of computational time. Using the computationally efficient procedure developed in this work, Aghyad (2014) carried out a parametric study to compare stationary and scanning laser bending processes.

4.10   Inverse modeling, optimization, and control

Laser bending has drawn the attention of a number of researchers and engineers. Various numerical and analytical models are developed to predict and to understand insight of the process. However, one major difficulty in using these models is that many times some properties of workpiece materials, properties of laser beam, and properties related to laser-to-work piece interaction are not known. In this scenario, use of the inverse modeling is important for efficient and accurate modeling of the laser bending process. Also, various process parameters have nonlinear effects on bend angle and edge effect and therefore, the process planning and control of irradiation strategies and process conditions are important for optimization of the laser bending process. In the following subsections these aspects are discussed briefly.

4.10.1   Inverse modeling and optimization

In inverse modeling, a suitable optimization technique is used to estimate the unknown parameters when the experimental/known results are sufficiently closer to those predicted by the direct models. The results obtained from the analytical or numerical model are compared with experimental or known data. Based on the difference, values of unknown parameters are adjusted. The values of unknown parameters are optimized by minimizing suitable error function. Mishra and Dixit (2013) determined absorptivity, thermal diffusivity, and laser beam diameter by an inverse heat conduction method. They measured temperature at centroid on the bottom surface at different time intervals. The thermal properties were estimated from the temperature measurements. The minimization of objective function was based on absorptivity, thermal diffusivity, and beam diameter. Figure 4.17 shows the comparison of experimental temperature variation along with the temperature variation predicted by the inverse model. The methodology was found to be efficient. However, more than one combination of parameters can provide the same temperature variation with time at a particular location. Therefore, Aghyad and Dixit (2013) improved this model and they carried out inverse determination by measuring temperature at two locations. The objective was to minimize combined errors between predicted and measured temperature at two locations. The proposed methodology was found to be efficient and robust.

The optimization of the laser bending process is one promising area. There is hardly any comprehensive research on the optimization of laser bending. There are some papers

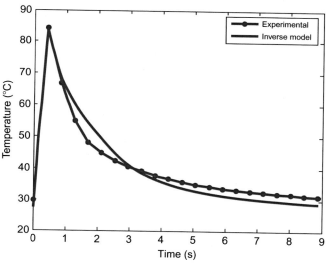

Figure 4.17 Variation of temperature with time (Mishra and Dixit, 2013). Copyright Nova Science Publishers.

on the application of soft computing for predicting the bend angle. Kumar and Dixit (2008) combined a finite element module with the artificial intelligence module containing a neural network that learns with experience. The inverse problem of determining the traverse speed for the required bend angle is solved using an optimization module. With the help of numerical experiments, the effectiveness of the proposed scheme is demonstrated. The scheme is helpful in reducing the computational time considerably.

4.10.2 Process design and control

Laser bending can be used to generate various complex shapes using different irradiation strategies. The laser parameters such as laser power, scanning velocity, beam diameter, and scanning path should be optimized to get the desired shape. Many methodologies have been proposed by researchers to use feedback provided by the bent specimen to alter the irradiation strategies. These methodologies were able to get the desired shape using laser beam irradiations. Liu, Yao, and Srinivasan (2004) proposed an optimization approach for determination of optimal laser scanning paths and heating conditions to produce doubly curved shapes. The approach includes strain field calculation based on principal curvature formulation and minimal strain optimization. The scanning paths and heating condition (laser power and scanning velocity) are determined by combining analytical and practical constraints. Later, Liu and Yao (2005) developed the FEM-based design methodology to determine the laser scanning paths and heating conditions. The strain fields were first calculated by using FEM and decomposed into in-plane and bending strains. The laser scanning paths were chosen perpendicular to the minimum principal strain direction. The heating conditions were decided based on the ratios of in-plane to the bending strain.

Kim and Na (2003) proposed distance-based and angle-based criteria algorithms to generate laser scanning paths. The distance-based method used a maximum distance between the given sheet and the target shape as a criterion for a new irradiation point. When the maximum distance was larger than the offset distance, the point at the maximum distance was adopted as a new irradiation point. The second method used angles between tangent lines as a criterion for making an irradiation point. The angle gradually increases as the radius of the curvature decreases, so that more forming points can be generated in a highly curved surface. Kim and Na (2005) employed a feedback control scheme for 2-D laser curve bending. The feedback was obtained by means of laser displacement sensors. This scheme improved the accuracy of the final product.

Researchers proposed some methodologies to generate complex shapes like pillow, saddle, and sine wave. However, improvement in their accuracy and surface quality is still a challenging task. In the near future, this area is expected to gain much importance.

4.11   Conclusion

The research on the laser sheet bending process is gaining importance as is evident from the large number of papers appearing in journals. In the last decade, a number of researchers developed analytical or numerical models of the process. The theoretical modeling of laser bending involves the transient thermal as well as elasto-plastic analysis. As such, the solution by numerical methods has to be done time-increment-wise. The problem is compounded because of the unavailability of the temperature dependent thermal and mechanical properties. This problem can be obviated by developing the inverse methods. Soft computing techniques can also help in efficient industrial application of the process.

The challenging research areas in laser sheet bending are as follows:

- Development of efficient variants of the process such as laser bending with a moving mechanical load or rapid cooling after each pass.
- Development of computationally efficient mathematical models for the process.
- Development of suitable inverse techniques for extracting the time-dependent mechanical and thermal properties, laser parameters, and laser sheet interaction parameters.
- Process design and control.
- Studies on microstructure and mechanical properties of the product.

It is also envisaged that in the future miniaturized laser systems will be developed for microforming.

Acknowledgments

The authors are grateful to Prof. J. Paulo Davim for inviting them to submit this article. Some figures in this article have been taken from published sources for the purpose of review. Authors acknowledge the following authors and publishers: Sage Publications, Nova Science Publishers, Elsevier, AIMTDR Conference 2012, IIT Guwahati, IIT (BHU), Mr. Aghyad Eideh, Mr. P. Dutta, Mr. A.K. Ray, Mr. K. Singh, and Dr. K. Kalita.

References

Aghyad, A. (2014). *Determination of parameters during laser bending by inverse analysis.* M.Tech. thesis. India: Department of Mechanical Engineering, IIT Guwahati.

Aghyad, A., & Dixit, U. S. (2013). A robust and efficient inverse method for determining the thermal parameters during laser forming. In: *Proceedings of national conference of recent advancements in mechanical engineering*, November 8–9. Nirjuli, India: NERIST (pp. 38–43).

Chen, D., Wu, S., & Li, M. 2004a. Studies on laser forming of Ti–6Al–4V alloy sheet. *Journal of Materials Processing Technology, 152*(1), 62–65.

Chen, D., Wu, S., & Li, M. 2004b. Deformation behaviours of laser curve bending of sheet metals. *Journal of Materials Processing Technology, 148*(1), 30–34.

Chen, G., & Xu, X. (2001). Experimental and 3D finite element studies of CW laser forming of thin stainless steel sheets. *Journal of Manufacturing Science and Engineering, 123*, 66–73.

Chen, G., Xu, X., Poon, C. C., & Tam, A. C. (1999). Experimental and numerical studies on microscale bending of stainless steel with pulsed laser. *Transaction of the ASME Journal of Applied Mechanics, 66*, 772–779.

Cheng, P., Fan, Y., Zhang, J., Mika, D. P., Zhang, W., Graham, M., et al. (2006). Laser forming of varying thickness plate—Part 1: Process analysis. *ASME Journal of Manufacturing Science and Engineering, 128*, 634–641.

Cheng, J., & Yao, Y. (2001). Cooling effects in multiscan laser forming. *Journal of Manufacturing Processes, 3*(1), 60–72.

Duflou, J., Callebaut, B., Verbert, J., & Baerdemaeker, H. (2007). Laser assisted incremental forming: Formability and accuracy improvement. *Annals of the CIRP, 56*(1), 273–276.

Dutta, P. P., Kalita, K., & Dixit, U. S. (2013). Experimental investigation on laser bending of mild steel coated with black enamel paint. In: *Proceedings of national conference on manufacturing: Vision for future*, October 12–13, 2013. India: IIT Guwahati (pp. 198–203).

Edwardson, S., Abed, E., Bartkowiak, K., Dearden, G., & Watkins, K. (2006). Geometrical influences on multi-pass laser forming. *Journal of Physics D: Applied Physics, 39*(2), 382–389.

Fan, Y., Yang, Z., Cheng, P., Egland, K., & Yao, L. (2005). Investigation of effect of phase transformations on mechanical behavior of AISI 1010 steel in laser forming. *Journal of Manufacturing Science and Engineering, 129*(1), 110–116.

Geiger, M., Vollertsen, F., & Deinzer, G. (1993). Flexible straightening of car body shells by laser forming. SAE Technical Paper 930279. doi: 10.4271/930279.

Gillner, A., Holtkamp, J., Hartmann, C., Olowinsky, A., Gedicke, J., Klages, K., et al. (2005). Laser applications in microtechnology. *Journal of Materials Processing Technology, 167*, 494–498.

Hu, Z., Wang, H., Labudovic, M., & Kovacevic, R. (2001). Computer simulation and experimental investigation of sheet metal bending using laser beam scanning. *International Journal of Machine Tools and Manufacture, 41*, 589–607.

Hu, J., Xu, H., & Dang, D. (2013). Modeling and reducing edge effects in laser bending. *Journal of Materials Processing Technology, 213*(11), 1989–1996.

Kant, R., & Joshi, S. N. (2012a). Numerical simulation of laser bending of magnesium alloy AZ31B using FEM. In: *Proceedings of the international conference IDDRG 2012*, held at IIT Bombay during November 25–28, Vol. 2 (pp. 736–741).

Kant, R., & Joshi, S. N. (2012b). Thermo-mechanical analysis of curvilinear laser bending of magnesium alloy sheet using finite element method. In: *4th international & 25th AIMTDR conference*, Jadavpur University, December 14–16, 2012 (pp. 109–114).

Kant, R., & Joshi, S. N. (2014). Numerical modeling and experimental validation of curvilinear laser bending of magnesium alloy sheets. *Proceedings of the Institution of Mechanical Engineers, Part B: Journal of Engineering Manufacture, 228,* 1036–1047. http://dx.doi.org/10.1177/0954405413506419.

Kant, R., & Joshi, S. N. (2013a). Finite element simulation of laser assisted bending with moving mechanical load. *International Journal of Mechatronics and Manufacturing Systems, 6*(4), 351–366.

Kant, R., & Joshi, S. N. (2013b). Numerical simulation of multi-pass laser bending processes using finite element method. In: *Proceedings of the 2nd international conference IRAM 2013,* held at IIT Indore during December 16–18 (pp. 208–213).

Kant, R., & Joshi, S. N. (2013c). Finite element simulation of multi-pass laser bending process with forced cooling. In: *Proceedings of the 8th international conference COPEN 2013,* held at NIT Calicut during December 13–15 (pp. 772–777).

Kant, R., Joshi, S. N., & Dixit, U. S. (2013a). State of the art and experimental investigation on edge effect in laser bending process. In: *Proceedings of the national conference NCRAME 2013,* held at NERIST, Itanagar during November 8–9 (pp. 189–197).

Kant, R., Joshi, S. N., & Dixit, U. S. (2013b). Experimental studies on laser bending of magnesium m1a alloy sheet. In: *Proceedings of the national conference MVF 2013,* held at IIT Guwahati during October 12–13 (pp. 189–197).

Kim, J., & Na, S. (2003). Development of irradiation strategies for free curve laser forming. *Optics and Laser Technology, 35*(18), 605–611.

Kim, J., & Na, S. (2005). Feedback control for 2D free curve laser forming. *Optics and Laser Technology, 37*(2), 139–146.

Kitamura, K. (1983). *Materials processing by high powered laser.* Technical report. JWESTP-8302 Japan Welding Engineering Society, pp. 359–371.

Kraus, J. (1997). Basic processes in laser bending of extrusions using the upsetting mechanism. In: *Laser assisted net shape engineering 2. Proceedings of the LANE'97,* Meisenbach, Bamberg.

Kumar, G. R. S., & Dixit, U. S. (2008). Determination of traverse speed in the laser forming by using FEM with online learning. In: *2nd international & 23rd all India manufacturing technology, design and research conference,* December 15–17. Chennai: IIT Madras.

Kyrsanidi, A. K., Kermanidis, T. B., & Pantelakis, S. G. (1999). Numerical and experimental investigation of the laser forming process. *Journal of Materials Processing Technology, 87,* 281–290.

Kyrsanidi, A. K., Kermanidis, T. B., & Pantelakis, S. G. (2000). An analytical model for the predictions of distortions caused by laser forming process. *Journal of Materials Processing Technology, 104,* 94–102.

Lambiase, F. (2012). An analytical model for evaluation of bending angle in laser forming of metal sheets. *Journal of Materials Engineering and Performance, 21,* 2044–2052.

Lambiase, F., & Ilio, A. D. (2013). A closed-form solution for thermal and deformation fields in laser bending process of different materials. *International Journal of Advanced Manufacturing Technology, 69,* 849–861. http://dx.doi.org/10.1007/s00170-013-5084-9.

Lawrence, J., Schmidt, M. J. J., & Li, L. (2001). The forming of mild steel plates with a 2.5 kW high power diode laser. *International Journal of Machine Tools & Manufacture, 41*(7), 967–977.

Li, W., & Yao, Y. (2001). Numerical and experimental investigation of convex laser forming process. *Journal of Manufacturing Processes, 3*(2), 73–81.

Liu, J., Sun, S., Guan, Y., & Ji, Z. (2010). Experimental study on negative laser bending process of steel foils. *Optics and Lasers in Engineering, 48*(1), 83–88.

Liu, C., & Yao, Y. (2005). FEM-based process design for laser forming of doubly curved shapes. *Journal of Manufacturing Processes, 7*(2), 109–121.

Liu, C., Yao, Y., & Srinivasan, V. (2004). Optimal process planning for laser forming of doubly curved shapes. *Journal of Manufacturing Science and Engineering, 126*(1), 1–9.

McGrath, P., & Hughes, C. (2007). Experimental fatigue performance of laser-formed components. *Optics and Lasers in Engineering, 45*(3), 423–430.

Mishra, A., & Dixit, U. S. (2013). Determination of thermal diffusivity of the material, absorptivity of the material and laser beam radius during laser forming by inverse heat transfer. *Journal of Machining and Forming Technologies, 5*(3–4), 207–226.

Ojeda, C., & Grej, J. (2009). Bending of stainless steel thin sheets by a raster scanned low power CO_2 laser. *Journal of Materials Processing Technology, 209*(5), 2641–2647.

Shen, H., Hu, J., & Yao, Z. (2010). Analysis and control of edge effects in laser bending. *Optics and Lasers in Engineering, 48*(3), 305–315.

Shen, H., Hu, J., & Yao, Z. (2011). Cooling effects in laser forming. *Materials Science Forum, 663–665*, 58–63.

Shen, H., & Yao, Z. (2009). Study on mechanical properties after laser forming. *Optics and Lasers in Engineering, 47*(1), 111–117.

Shen, H., Yao, Z., Shi, Y., & Hu, J. (2006). An analytical formula for estimating the bending angle by laser forming. *Proceedings of the Institution of Mechanical Engineering, Part C: Journal of Mechanical Engineering Science, 220*, 243–247.

Shi, Y., Liu, Y., Yao, Z., & Shen, H. (2008). A study on bending direction of sheet metal in laser forming. *Journal of Applied Physics, 103*, 053101.

Shi, Y., Shen, H., Yao, Z., & Hu, J. (2007). Temperature gradient mechanism in laser forming of thin plates. *Optics & Laser Technology, 39*, 858–863.

Shi, Y., Yao, Z., Shen, H., & Hu, J. (2006). Research on the mechanisms of laser forming for the metal plate. *International Journal of Machine Tools & Manufacture, 46*(12–13), 1689–1697.

Shichun, W., & Jinsong, Z. (2001). An experimental study of laser bending for sheet metals. *Journal of Materials Processing Technology, 110*(2), 160–163.

Singh, K. (2013). *Effect of lime coating on laser bending process.* M.Tech. thesis IIT Guwahati.

Singh, K., Joshi, S. N., Ray, A. K., & Dixit, U. S. (2013). A comparison of bend quality of mechanical and laser bending of mild steel. In: *Proceedings of national symposium on miniature manufacturing in 21st century (NSMMIC-2013)*, August 16–18, 2013.Varanasi, India: IIT (BHU).

Vollertsen, F. (1994). An analytical model for laser bending. *Laser in Engineering, 2*, 261–276.

Yao, Z., Shen, H., Shi, Y., & Hu, J. (2007). Numerical study on laser forming of metal plates with pre-loads. *Computational Materials Science, 40*(1), 27–32.

Yau, C. J., Chan, K. C., & Lee, W. B. (1997). A new analytical model for laser bending. In: *Laser assisted net shape engineering 2. Proceedings of the LANE'97*, Meisenbach, Bamberg.

Yilbas, B. S., Arif, A. F. M., & Abdul Aleem, B. J. (2012). Laser bending of AISI 304 steel sheets: Thermal stress analysis. *Optics & Laser Technology, 44*(2), 303–309.

Zahrani, E. G., & Marasi, A. (2013). Experimental investigation of edge effect and longitudinal distortion in laser bending process. *Optics & Laser Technology, 45*, 301–307.

Zhang, P., Guo, B., Shan, D., & Ji, Z. (2007). FE simulation of laser curve bending of sheet metals. *Journal of Materials Processing Technology, 184*(1–3), 157–162.

Multiple performance optimization in drilling using Taguchi method with utility and modified utility concepts

V.N. Gaitonde[1], S.R. Karnik[1], J. Paulo Davim[2]
[1]B.V.B. College of Engineering and Technology, Hubli, India; [2]University of Aveiro, Aveiro, Portugal

5.1 Introduction

Drilling is extensively used in the manufacturing of mechanical and electrical components. Burr is plastically deformed material generated on the edge of the component during drilling.

Wherever drilling is applied, the burr formation on a sheared edge is usual. Such burrs are produced instantly at the commencement of cutting when the chisel edge comes in contact with the workpiece. The burrs are formed and propagated in a circumferential direction as the drill is fed into the workpiece.

The precise definition of a burr, lack of understanding of burr formation, and implications of burr on quality during the drilling process in automotive and aerospace manufacturing society are the critical problems (Koelsch, 2001). The occurrence of burr on a component's edges reduces the quality of the product, poses difficulties in a machining sequence as well as assembly, and thus hinders the industrial automation. It has been reported that the burrs may cause blockage of critical passages and turbulence in the flow of liquids or gases through conduits, which might cause severe problems during service (Avila, Choi, Dornfeld, Kapgan, & Kosarchuk, 2004). The burrs can cause short circuits in electrical components, reduce the fatigue life of components, and act as a crack initiation point (Min, Kim, & Dornfeld, 2001a). In case of parts moving relative to each other, friction and wear due to burrs not only reduce the edge quality but also produce noise and vibration. In the computer storage manufacturing industry, an attached burr may inhibit assembly of the file or a detached burr may later restrict file function or cause disk crash (Bakkal, Shih, McSpaadden, Liu, & Scattergood, 2005). The exit burr has a bad influence on the ejection of a chip, accuracy of a finished product, surface roughness, and drill wear. In addition, a burr is a hindrance to all the succeeding processes. The exit burrs are injurious, as they hit the cutting edge and cause the groove wear, which in turn accelerates the burr growth (Arai & Nakayama, 1986). Hence, it is essential to remove the burrs from part's edges. The standard deburring tools are necessary for removing the exit burrs formed inside a cavity. The cost of deburring of components may contribute as much as 30% to the

Materials Forming and Machining. http://dx.doi.org/10.1016/B978-0-85709-483-4.00005-3

cost of finished parts and secondary finishing operations are found to be difficult to automate. Dornfeld (1992) reported that secondary processes used to remove the exit burrs are expensive and, therefore, there is a need to reduce the deburring effort.

The effects of process parameters on burr formation during drilling of various work materials have been studied by Dornfeld, Kim, Dechow, Hewson, and Chen (1999), Ko and Lee (2001), Lauderbaugh (2009), Lin (2002), and Stein and Dornfeld (1997). Furthermore, finite element models (Guo & Dornfeld, 2000; Min, Kim, & Dornfeld, 2001b; Saunders, 2003), empirical drilling charts (Kim & Dornfeld, 2000; Kim, Min, & Dornfeld, 2001; Min, Dornfeld, Kim & Shyu, 2001), response surface methodology-based mathematical models (Gaitonde, Karnik, Achyutha, Siddeswarappa, & Davim, 2009; Gaitonde, Karnik, Siddeswarappa, & Achyutha, 2008), and artificial neural network models (Gaitonde & Karnik, 2012a; Karnik & Gaitonde, 2008; Karnik, Gaitonde, & Davim, 2008) have been proposed for analyzing the burr formation in drilling. Modified Taguchi methods have been used for minimizing the burr size in drilling (Gaitonde & Karnik, 2007, 2012b; Gaitonde, Karnik, Achyutha, & Siddeswarappa, 2007, 2008; Hunag, 2004; Tosun, 2006). In the recent past, nontraditional heuristic search algorithms like genetic algorithms and particle swarm optimization have also been employed to minimize the burr size in drilling (Gaitonde & Karnik, 2012a; Gaitonde, Karnik, & Davim, 2012; Gaitonde, Karnik, Siddeswarappa, et al., 2008; Karnik, Gaitonde, & Davim, 2007).

The Taguchi robust design methodology has produced a prominent optimization tool that differs from that used in customary practices (Phadke, 1989; Ross, 1996). This methodology can reasonably satisfy the needs of problem solving and design optimization with a reduced number of experiments. The original Taguchi design was applied to optimize a single performance characteristic and employed in the past for optimization (Deng & Chin, 2005; Yang & Chen, 2001). On the contrary, most of the products/processes have numerous performance characteristics. Hence, there is a need for a single optimal process parameter setting with optimum or near optimum quality characteristics as a whole. Several modifications were suggested to the original Taguchi method for multiperformance characteristic optimization (Jeyapaul, Shahabudeen, & Krishnaiah, 2005).

The exit burr size is a performance indicator of the drilling process, which decides the quality of the finished products. Therefore, it is essential to minimize the burr formation at the manufacturing stage by selecting the appropriate drilling process parameters. This necessitates an appropriate optimization tool to minimize the burr size for a specified combination of drilling process parameters. This chapter presents the application of the Taguchi method with utility concept as well as modified utility concepts for multiobjective drilling process optimization to determine the optimal values of cutting speed, feed, point angle, and clearance angle to simultaneously minimize burr size, namely, burr height and burr thickness during drilling of AISI 304 stainless steel workpieces using HSS twist drills. AISI 304 stainless steel work materials find extensive applications in chemical industries. Conversely, burr formation during drilling is a key problem due to ductility of the stainless steel work material.

The utility concept proposed by Kumar, Barua, and Gaindhar (2000) employs the weighting factors to each of the signal-to-noise (S/N) ratios of the responses to obtain a multiresponse *S/N* ratio for each trial of the orthogonal array (OA). The present investigation introduces a new modification to the utility concept in addition to the utility concept proposed by Kumar et al. (2000). This new modification employs weighting factors directly to the responses to obtain the multiresponse objective function. The effectiveness of utility concept, the modified utility concept, and the effect of weighting factors on optimization results are analyzed in detail in this chapter.

5.2 Taguchi method with utility concept

The Taguchi method is used for finding the best combinations of the control factors to make the product or process insensitive to the noise factors (Phadke, 1989; Ross, 1996). The Taguchi method is based upon the technique of matrix experiments. Experimental matrices are special OAs (Phadke, 1989; Ross, 1996) which allow the concurrent effect of numerous process parameters to be studied capably. The purpose of conducting an orthogonal experiment is to determine the optimum level for each parameter and to establish the comparative significance of individual factors in terms of direct effects on the response. Taguchi suggests the *S/N* ratio as the objective function for matrix experiments (Phadke, 1989; Ross, 1996). The *S/N* ratio is used to measure the quality characteristics and the significant process parameters through analysis of variance (ANOVA). Taguchi classifies the objective functions into three categories: namely, the smaller the better type, the larger the better type, and the nominal the best type. The optimum level for a factor is the level that results in the highest value of the *S/N* ratio.

In the present investigation, the Taguchi robust design technique is employed to optimize the drilling process. This requires determination of optimal values of cutting speed (v), feed (f), point angle (θ), and clearance angle (ψ) to simultaneously minimizing burr height (B_h) and burr thickness (B_t) during drilling of AISI SS304 work material. Thus, the multiresponse drill optimization problem can be stated as:

Determine v, f, θ, and ψ.
So as to minimize B_h and B_t.
Subject to constraints $X_{min} \leq X \leq X_{max}$ for $X = v, f, \theta$, and ψ.

Because, the original Taguchi technique was designed to optimize a single performance characteristic, the method requires modification for multiresponse optimization. The simplest modification, known as utility concept, employs weighting factors to each of the *S/N* ratio of responses to obtain the multiresponse *S/N* ratio for each trial of the OA. The *S/N* ratio associated with the responses B_h and B_t are given as:

$$\eta_1 = -10 \log_{10} \left(B_h^2 \right) \tag{5.1}$$

$$\eta_2 = -10 \log_{10} \left(B_t^2 \right) \tag{5.2}$$

In the utility concept, the multiresponse *S/N* ratio is defined as:

$$\eta = w_1 \eta_1 + w_2 \eta_2 \tag{5.3}$$

where w_1 and w_2 are the weighting factors associated with S/N ratio of responses B_h and B_t, respectively. These weighting factors are decided based on the priorities among various responses to be simultaneously optimized.

The above form of utility concept depends on the weighting factors applied to the S/N ratios. The present investigation also introduces a new modification to the utility concept. This new modification employs weighting factors directly to the responses to obtain the multiresponse objective function (Ob_{mr}) and is given as:

$$Ob_{mr} = w_1 B_h + w_2 B_t \tag{5.4}$$

Thus, the S/N ratio associated with each trial of L_9 OA is given as:

$$\eta = -10 \log_{10}\left[\left(w_1 B_h + w_2 B_t \right)^2 \right] \tag{5.5}$$

The effectiveness of utility and modified utility concepts, the effect of weighting factors on the optimization results, are analyzed and presented in the subsequent sections.

5.3   Experimental procedure

5.3.1   Orthogonal array

The traditional methods of experimentation are too complex and time consuming. A huge number of experiments are to be performed for a large number of parameters. Hence, the Taguchi method uses a special design of OAs to study the whole parameter space with merely a small number of experiments.

In the current investigation, four parameters; namely, cutting speed (v), feed (f), point angle (θ), and clearance angle (ψ) were identified as the control factors. The ranges for cutting speed, feed, and clearance angle were determined through preliminary experiments. Stein (1997), in her investigation, recommended a higher point angle for drilling of stainless steel work material and hence a point angle in the range 118–134° has been selected. One of the most universal tool materials for drilling stainless steel work material is HSS (Kim et al., 2001) and hence the HSS twist drill was selected in the current study. In the present investigation, each parameter was investigated at three levels to study the nonlinearity effects of the parameters. The identified parameters and their levels are summarized in Table 5.1.

The Taguchi optimization process begins with the selection of an OA with a distinct number of levels (L) defined for each of the parameters. The minimum number of trials in the OA is

$$N_{min} = \left(L - 1 \right) F + 1 \tag{5.6}$$

where F is the number of factors $= 4$ and L is the number of levels $= 3$.

This gives $N_{min} = 9$, and hence according to the Taguchi design (Phadke, 1989; Ross, 1996), an L_9 OA with four columns and nine rows was selected. Each process

Table 5.1 **Drilling process parameters and their levels**

Parameter	Code	Notation	Levels		
Cutting speed (m/min)	A	v	8	12	16
Feed (mm/rev)	B	f	0.04	0.12	0.20
Point angle (°)	C	θ	118	126	134
Clearance angle (°)	D	ψ	8	10	12

parameter is assigned to a column and only nine experiments are required to study the complete drilling process parameter space. The L_9 OA for the current investigation is shown in Table 5.2.

5.3.2 *Experimentation and exit burr size measurement*

The work material used throughout the study was AISI 304 stainless steel; the chemical composition and mechanical properties of the material are given in Table 5.3. The drilling operation was performed using precision drill check, HSS twist drill of 10 mm diameter with 30° helix angle (Addison & Co. Ltd., India). The drilling experiments were performed as per L_9 OA on three-axis "*YCM-V116*" CNC vertical machining center (VMC) (Yeong Chin Machinery Industries Co., Taiwan) with Fanuc controller. The VMC was equipped with a 15 kW drive motor, maximum feed of 5000 mm/min, and spindle speed from 45 to 4000 rpm. Initially, all work pieces were premachined on a milling machine to achieve parallelism within 50 μm. Test specimens of 10 mm thickness were polished on both sides and were placed on the fixture of VMC such that the drill bit could pass through both surfaces. Finally, drilling was carried out for nine different process parameter combinations of OA with Cut 60 EP as water-soluble coolant.

Table 5.2 L_9 **orthogonal array along with measured values of burr height and burr thickness**

	Levels of input parameters				Measured values of burr size	
Trial no.	A	B	C	D	Burr height (mm)	Burr thickness (mm)
1	1	1	1	1	1.44	0.016
2	1	2	2	2	0.52	0.118
3	1	3	3	3	0.785	0.009
4	2	1	2	3	0.618	0.065
5	2	2	3	1	0.915	0.043
6	2	3	1	2	0.54	0.049
7	3	1	3	2	1.037	0.046
8	3	2	1	3	0.756	0.024
9	3	3	2	1	1.605	0.056

Table 5.3 Chemical composition and mechanical properties of AISI 304 stainless steel work material

Chemical composition (wt%)	0.08 C, 18–20 Cr, 66–74 Fe, 2 Mn, 8–10 Ni, 0.045 P, 0.03 S, 1.0 Si
Mechanical properties	Yield strength: 215 MPa Ultimate tensile strength: 505 MPa Hardness: 123 BHN Modulus of elasticity: 193–200 GPa

Burr height and thickness were measured using "*RPP-400*" toolmakers' microscope (Sicherun-Gen Versehen, Germany) with a resolution of 1 μm and 30× magnification.

The average of the burr measurements along two diameters at right angles to each other was taken for the burr thickness of each hole. Measurements of burr height at 90° intervals around the hole were averaged to yield the average burr height. The measured burr size values as per L_9 OA are presented in Table 5.2.

5.4   Results and discussion

In the current study, three cases of different combinations of weighting factors are assumed. For case 1, the order of response is $w_1 = 0.9$ for burr height and $w_2 = 0.1$ for burr thickness, in other words, giving importance to burr height. In case 2, the weighting factors of 0.5 are considered for each of the responses, which give equal priorities to both burr height and burr thickness in the drilling process optimization. In the third case, the order of performance characteristics is $w_1 = 0.1$ for burr height, and $w_2 = 0.9$ for burr thickness, in other words, putting more prominence on burr thickness.

5.4.1   Taguchi optimization using utility concept

The multiresponse *S/N* ratio (η) for different combinations of the weighting factors using utility concept are computed using Equation (5.3) and are illustrated in Table 5.4. The analysis of means (ANOM) is used to determine the optimal levels of the process parameters and the effect of a factor level is the deviation it causes from the overall mean response (Phadke, 1989; Ross, 1996). The optimum level of a factor is the level that gives the highest *S/N* ratio. The results of ANOM for the three different cases considered are presented in Figures 5.1–5.3. Thus, the optimal process parameter levels for case 1, with more emphasis on burr height, is A2, B2, C1, and D3; for case 2 that gives equal importance to burr height and burr thickness is A1, B3, C1, and D3; and for case 3 is A1, B3, C3, and D3 that gives more importance to burr thickness. The results of optimization for different combinations of weighting factors are summarized in Table 5.5.

In all three cases considered above, the weighting factor w_1 is decreased while simultaneously increasing w_2 such that $w_1 + w_2 = 1.0$. This transfers the emphasis from burr height to burr thickness at an equal value of 0.5. Thus, with the increase in w_2 value (or with decrease of w_1), the optimal value of burr thickness is expected to reduce

Table 5.4 **Computed values of the multiresponse** *S/N* **ratio with different weighting factors**

Trial no.	Multiresponse *S/N* ratio, η (dB)					
	Utility concept			Modified utility concept		
	Case 1: $w_1=0.9$ $w_2=0.1$	Case 2: $w_1=0.5$ $w_2=0.5$	Case 3: $w_1=0.1$ $w_2=0.9$	Case 1: $w_1=0.9$ $w_2=0.1$	Case 2: $w_1=0.5$ $w_2=0.5$	Case 3: $w_1=0.1$ $w_2=0.9$
1	0.74123	16.37518	32.00912	2.75737	−2.26282	16.00490
2	6.96817	12.12115	17.27412	9.92418	6.378795	16.01587
3	5.98386	21.50888	37.0339	8.02419	3.006699	21.24964
4	6.13638	13.96098	21.78558	9.33218	4.994462	18.39469
5	3.42748	14.0511	24.67473	6.39329	1.641492	17.70778
6	7.43652	15.7741	24.11168	10.6182	6.180139	20.16662
7	2.39046	13.21463	24.0388	5.32803	0.556869	16.76665
8	5.42618	17.41267	29.39915	8.17870	3.31413	20.24667
9	−1.19492	10.46337	22.12167	1.61320	−3.22796	13.51847

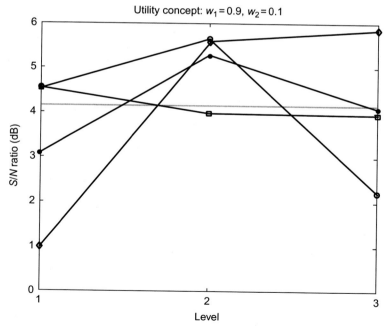

Figure 5.1 Response diagram of *S/N* ratio using utility concept for burr size ($w_1=0.9$ and $w_2=0.1$).

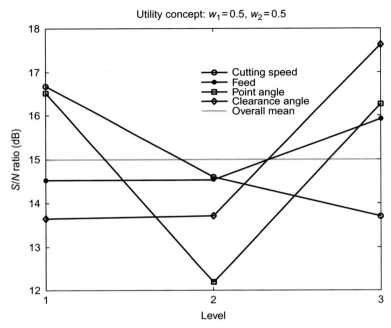

Figure 5.2 Response diagram of *S/N* ratio using utility concept for burr size (w_1=0.5 and w_2=0.5).

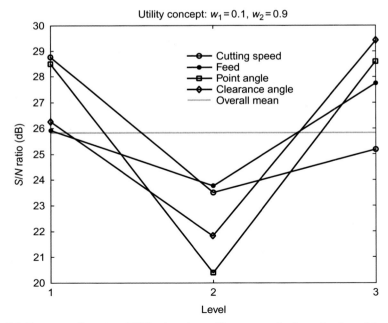

Figure 5.3 Response diagram of *S/N* ratio using utility concept for burr size (w_1=0.1 and w_2=0.9).

Table 5.5 **Optimization results for burr size**

| Case | Weighing factors | Utility concept | | | | | | Modified utility concept | | | | | | |
| | | Optimal level | | | Optimal burr size | | Optimal level | | | | Optimal burr size | |
		v	f	θ	ψ	B_h (mm)	B_t (mm)	v	f	θ	ψ	B_h (mm)	B_t (mm)
1	$w_1 = 0.9,$ $w_2 = 0.1$	2	2	1	3	0.661	0.045	2	2	2	2	0.248	0.112
2	$w_1 = 0.5,$ $w_2 = 0.5$	1	3	1	3	1.595	0.027	2	2	1	2	0.294	0.073
3	$w_1 = 0.1,$ $w_2 = 0.9$	1	3	3	3	0.785	0.009	2	3	1	3	0.841	0.007

while the optimal burr height increases. It is observed from Table 5.5 that the optimal value of burr thickness is decreasing with the increased value of weighting factor w_2. However, the optimal burr height does not follow the expected trend. Therefore, the weighting factors employed in the original utility concept do not really give actual emphasis on the various objectives of the optimization problem. In view of the above, the Taguchi method with a modified utility concept has been proposed in the following section.

5.4.2   Taguchi optimization using modified utility concept

The Taguchi technique with a modified utility concept is a new modification to the utility concept: it employs weighting factors directly to the responses to obtain the multiresponse objective function and S/N ratio associated with each trial of L_9 OA as given by Equation (5.5). The computed value of the multiresponse S/N ratio (η) for each trial of OA for different combinations of the weighting factors using the modified utility concept is given in Table 5.4. The results of ANOM for three different cases using the modified utility concept are displayed in Figures 5.4–5.6. The results of the Taguchi optimization using the modified utility concept for different combinations of weighting factors are summarized in Table 5.5.

It is observed from Table 5.5 that the optimal burr height and burr thickness obtained by the modified utility concept exactly follow the expected trend. Thus, the weighting factors associated with the various responses of the multiresponse

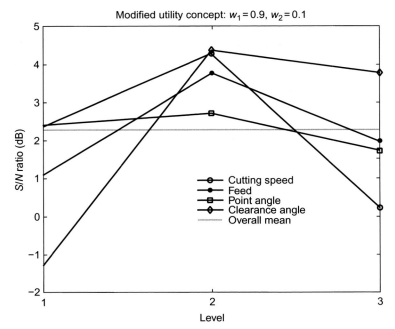

Figure 5.4 Response diagram of S/N ratio using modified utility concept for burr size ($w_1 = 0.9$ and $w_2 = 0.1$).

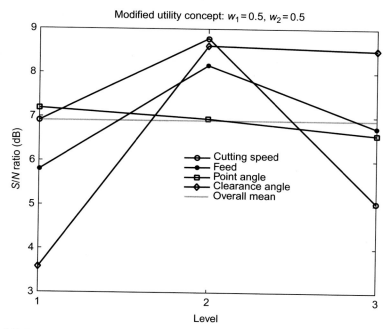

Figure 5.5 Response diagram of *S/N* ratio using modified utility concept for burr size ($w_1 = 0.5$ and $w_2 = 0.5$).

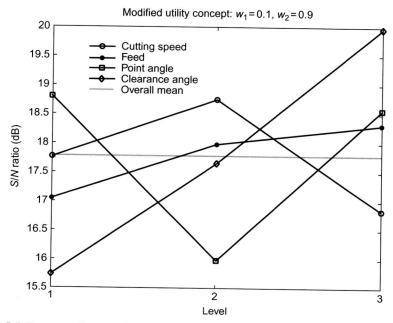

Figure 5.6 Response diagram of *S/N* ratio using modified utility concept for burr size ($w_1 = 0.1$ and $w_2 = 0.9$).

optimization problem is correctly mapped so as to give emphasis on the required response. Moreover, the optimum burr height and burr thickness obtained from the modified utility concept is much lower than those obtained from the utility concept.

From the above discussion, it is clear that the Taguchi optimization technique using the modified utility concept has yielded better results as compared to the utility concept when applied to a multiresponse drilling problem.

5.4.3 Analysis of variance

The relative importance among the parameters for multiple performance characteristics is further investigated by ANOVA. ANOVA is performed on the S/N ratio to obtain the contribution of each of the process parameters (Phadke, 1989; Ross, 1996). The results of ANOVA for the utility and modified utility concepts are given in Tables 5.6 and 5.7, respectively, for all three cases considered.

From Tables 5.6 and 5.7, it is observed that the contribution of each factor in controlling the burr size depends on the relative emphasis attached to burr height and burr thickness. When the emphasis is on minimizing burr height, the contributions from both cutting speed and clearance angle are much higher. On the other hand, when the emphasis is on minimizing burr thickness, larger contributions come from clearance angle and point angle. This clearly illustrates the importance of point angle in minimizing both burr height and burr thickness.

In addition, feed with contribution in the range 5–11% has a limited role in controlling burr. A medium to high value of feed can be employed in the drilling process as seen from Table 5.5. Medium values of feed always ensure lower thrust force, which reduces the burr formation. The thrust force also determines the amount of material that experiences plastic deformation during the final stage of the drilling process. Furthermore, the work material easily deforms and flows more easily at higher temperatures.

In our investigation, drilling of 10 mm diameter, medium values of cutting speed (12 m/min), point angle (126°), and clearance angle (10°) are found to be useful in minimizing both burr height and thickness with equal emphasis (Table 5.5). The effect of increased burr size with increasing feed is more pronounced at high cutting speeds where the temperature is higher. Hence, a combination of medium values of cutting speed and feed are preferred to minimize burr size. In general, medium to higher point angle assures maximum lip movement in the earliest possible time to avoid work hardening, resulting in thinner burrs due to change in the chip flow direction. Furthermore, for the given drill diameter, the plastic deformation is localized along the peripheral part of the hole, thus producing uniform and thinner burrs at a higher point angle. Low to medium clearance angle avoids weakening of cutting by providing support behind the cutting edges thus resulting in thinner burrs.

5.4.4 Confirmation tests

Once the optimal level of the parameters has been selected, the final step is to predict and verify the response using these parameters (Phadke, 1989). The experiments were conducted with the optimal levels of the parameters obtained by the Taguchi

Table 5.6 Analysis of variance (ANOVA) for burr size using utility concept

Parameter	Degrees of freedom (DF)	Case 1			Case 2			Case 3		
		Sum of squares (SS)	Mean square (MS)	Percent contribution	Sum of squares (SS)	Mean square (MS)	Percent contribution	Sum of squares (SS)	Mean square (MS)	Percent contribution
Cutting speed	2	18.7400	9.3700	26.21	13.9345	6.9673	16.45	43.1677	21.5839	15.03
Feed	2	7.1813	3.5907	10.05	3.8802	1.9401	4.58	23.7395	11.8698	8.26
Point angle	2	0.6811	0.3406	0.95	35.5110	17.7555	41.92	132.8789	66.4394	46.26
Clearance angle	2	44.8840	22.4420	62.79	31.3859	15.6929	37.05	87.4693	43.7346	30.45
Error	0	0	–	–	0	–	–	0	–	–
Total	8	71.4864	–	100	84.7116	–	100	287.2554	–	100

Table 5.7 ANOVA for burr size using modified utility concept

Parameter	Degrees of freedom (DF)	Case 1			Case 2			Case 3		
		Sum of squares (SS)	Mean square (MS)	Percent contribution	Sum of squares (SS)	Mean square (MS)	Percent contribution	Sum of squares (SS)	Mean square (MS)	Percent contribution
Cutting speed	2	24.7315	12.3658	25.94	20.9959	10.4979	26.37	5.4898	2.7449	11.05
Feed	2	11.1960	5.5980	11.75	8.4603	4.2302	10.63	2.5549	1.2775	5.15
Point angle	2	1.5096	0.7548	1.58	0.5561	0.2780	0.70	14.8123	7.4061	29.83
Clearance angle	2	57.8912	28.9456	60.73	49.6124	24.8062	62.30	26.7952	13.3976	53.97
Error	0	0	–	–	0	–	–	0	–	–
Total	8	95.3283	–	100	79.6247	–	100		–	100

optimization for different combinations of the weighting factors using both the utility and modified utility concepts. The measured values of burr height and burr thickness were used to determine the observed values of multiperformance objectives and the corresponding S/N ratio (η_{obs}). The predicted optimum value of the S/N ratio (η_{opt}) was then determined. It is found that the prediction error (i.e., the difference between η_{opt} and η_{obs}) is within the confidence interval value, clearly indicating the adequacy of the burr size models.

5.5 Conclusions

The methodology of the Taguchi optimization for the multiobjective drilling problem to minimize burr size is presented in this chapter. The approach is based on the utility concept, which employs weighting factors. In the original utility concept, the weighting factors are applied to the S/N ratio associated with each of the responses. Because this method does not emphasize the required responses of a multiresponse optimization problem, a new modification was introduced. The modified utility concept employs the weighting factors directly to the various responses to obtain the multiresponse objective function. The effectiveness of the approach is demonstrated through detailed analysis to determine the optimal process setting levels, which simultaneously minimize burr height and burr thickness during drilling of AISI 304 stainless steel workpieces. Optimal process parameters were confirmed with verification experiments.

References

Arai, M., & Nakayama, K. (1986). Boundary notch on cutting tool caused by burr and its suppression. *Bulletin of the Japan Society of Precision Engineering, 52,* 864–866.

Avila, M. C., Choi, J., Dornfeld, D. A., Kapgan, M., & Kosarchuk, R. (2004). *Deburring of cross-drilled hole intersections by mechanized cutting.* LMA annual research reports, pp. 10–20.

Bakkal, M., Shih, A. J., McSpaadden, S. B., Liu, C. T., & Scattergood, R. O. (2005). Light emission, chip morphology and burr formation in drilling the bulk metallic glass. *International Journal of Machine Tools and Manufacture, 45,* 741–752.

Deng, C. S., & Chin, J. H. (2005). Hole roundness in deep hole drilling as analysed by Taguchi methods. *The International Journal of Advanced Manufacturing Technology, 25,* 420–426.

Dornfeld, D. A. (1992). *Intelligent deburring of precision components. Proceedings of the international conference on industrial electronics, control, instrumentation and automation: Vol. 2* (pp. 953–960). San Diego: IEEE.

Dornfeld, D. A., Kim, J., Dechow, H., Hewson, J., & Chen, I. J. (1999). Drilling burr formation in titanium alloy, Ti–6Al–4V. *Annals of CIRP, 52*(1), 45–48.

Gaitonde, V. N., & Karnik, S. R. (2007). Taguchi robust design for multi-response drilling optimization to minimize burr size using utility concept. *International Journal of Manufacturing Research, 2*(2), 209–224.

Gaitonde, V. N., & Karnik, S. R. 2012a. Minimizing burr size in drilling using artificial neural network (ANN)-particle swarm optimization (PSO) approach. *Journal of Intelligent Manufacturing Systems, 23*(5), 1783–1793.

Gaitonde, V. N., & Karnik, S. R. 2012b. Selection of optimal process parameters for minimizing burr size in drilling using Taguchi's quality loss function approach. *Journal of the Brazilian Society of Mechanical Sciences & Engineering, 34*(3), 238–245.

Gaitonde, V. N., Karnik, S. R., Achyutha, B. T., & Siddeswarappa, B. (2007). Methodology of Taguchi optimization for multi-objective drilling problem to minimize burr size. *International Journal of Advanced Manufacturing Technology, 34*, 1–8.

Gaitonde, V. N., Karnik, S. R., Achyutha, B. T., & Siddeswarappa, B. (2008). Taguchi optimization in drilling of AISI 316L stainless steel to minimize burr size using multi-performance objective based on membership function. *Journal of Materials Processing Technology, 202*, 374–379.

Gaitonde, V. N., Karnik, S. R., Achyutha, B. T., Siddeswarappa, B., & Davim, J. P. (2009). Predicting burr size in drilling of AISI 316L stainless steel using response surface analysis. *International Journal of Materials and Product Technology, 35*(1–2), 228–245.

Gaitonde, V. N., Karnik, S. R., & Davim, J. P. (2012). Minimizing burr size in drilling: Integrating response surface methodology with particle swarm optimization. In J. P. Davim (Ed.), *Mechatronics and manufacturing—Research and development* (pp. 259–292). UK: Wood Head Publishing (Chapter 7).

Gaitonde, V. N., Karnik, S. R., Siddeswarappa, B., & Achyutha, B. T. (2008). Integrating Box–Behnken design with genetic algorithm to determine the optimal parametric combination for minimizing burr size in drilling of AISI 316L stainless steel. *International Journal of Advanced Manufacturing Technology, 37*(3–4), 230–240.

Guo, Y. B., & Dornfeld, D. A. (2000). Finite element modeling of drilling burr formation process in drilling 304 stainless steel. *ASME Journal of Manufacturing Science and Engineering, 122*(4), 612–619.

Hunag, M. F. (2004). Application of Grey–Taguchi method to optimize drilling of aluminium alloy 6061 with multiple performance characteristics. *Materials Science and Technology, 20*, 528–532.

Jeyapaul, R., Shahabudeen, P., & Krishnaiah, K. (2005). Quality management research by considering multi-response problems in the Taguchi method—A review. *The International Journal of Advanced Manufacturing Technology, 26*, 1331–1337.

Karnik, S. R., & Gaitonde, V. N. (2008). Development of artificial neural network models to study the effect of process parameters on burr size in drilling. *International Journal of Advanced Manufacturing Technology, 39*(5–6), 439–453.

Karnik, S. R., Gaitonde, V. N., & Davim, J. P. (2007). Integrating Taguchi principle with GA to minimize burr size in drilling of AISI 316L stainless steel using ANN model. *Proceedings of the Institute of Mechanical Engineering, IMech E, Journal of Engineering Manufacture, 221*, 1695–1704.

Karnik, S. R., Gaitonde, V. N., & Davim, J. P. (2008). A comparative study of the ANN and RSM modeling approaches for predicting burr size in drilling. *International Journal of Advanced Manufacturing Technology, 38*(9–10), 868–883.

Kim, J., & Dornfeld, D. A. (2000). Development of a drilling burr control chart for low alloy steel, AISI 4118. *Journal of Materials Processing Technology, 113*, 4–9.

Kim, J., Min, S., & Dornfeld, D. A. (2001). Optimization and control of drilling burr formation of AISI 304L and AISI 4118 based on drilling burr control charts. *International Journal of Machine Tools and Manufacture, 41*, 923–936.

Ko, S. L., & Lee, J. (2001). Analysis on burr formation in drilling with new concept drill. *Journal of Materials Processing Technology, 113*, 392–398.

Koelsch, J. (2001). Divining edge quality by reading the burrs. *Quality Magazine*, 24–28.

Kumar, P., Barua, P. B., & Gaindhar, J. L. (2000). Quality optimization (multi-characteristic) through Taguchi's technique and utility concept. *Quality and Reliability Engineering International, 16*(6), 475–485.

Lauderbaugh, L. K. (2009). Analysis of the effects of process parameters on exit burrs in drilling using a combined simulation and experimental approach. *Journal of Materials Processing Technology, 209,* 1909–1919.

Lin, R. (2002). Cutting behavior of a TiN-coated carbide drill with curved cutting edges during the high speed machining of stainless steel. *Journal of Materials Processing Technology, 27,* 8–16.

Min, S., Dornfeld, D. A., Kim, J., & Shyu, B. (2001). Finite element modeling of burr formation in metal cutting. *Machining Science and Technology, 5*(2), 307–322.

Min, S., Kim, J., & Dornfeld, D. A. (2001a). Development of a drilling burr control chart for low alloy steel, AISI 4118. *Journal of Materials Processing Technology, 113,* 4–9.

Min, S., Kim, J., & Dornfeld, D. A. (2001b). Development of a drilling burr control chart for stainless steel. *Transactions of NAMRI/SME, 28,* 317–322.

Phadke, M. S. (1989). *Quality engineering using robust design.* Englewood Cliffs, NJ: Prentice Hall.

Ross, P. J. (1996). *Taguchi techniques for quality engineering.* New York: McGraw-Hill.

Saunders, L. K. L. (2003). A finite element modeling of exit burrs for drilling of metal. *Finite Element in Analysis and Design, 40*(2), 139–158.

Stein, J. M. (1997). *The burrs from drilling: An introduction to drilling burr technology.* USA: Burr Technology Information Series™.

Stein, J. M., & Dornfeld, D. A. (1997). Burr formation in drilling miniature holes. *Annals of CIRP, 46*(1), 63–66.

Tosun, N. (2006). Determination of optimum parameters for multi-performance characteristics in drilling by using grey relational analysis. *International Journal of Advanced Manufacturing Technology, 28,* 450–455.

Yang, J. L., & Chen, J. C. (2001). A systematic approach for identifying optimum surface roughness performance in end milling operations. *Journal of Industrial Technology, 17*(2), 1–8.

Molecular dynamics simulation of material removal with the use of laser beam

A.P. Markopoulos[1], P. Koralli[1,2], G. Kyriakakis[1], M. Kompitsas[2], D.E. Manolakos[1]
[1]National Technical University of Athens (NTUA), Athens, Greece; [2]Theoretical and Physical Chemistry Institute (TPCI), Athens, Greece

6.1 Introduction

Over the last 20 years, laser technology has often been cited in micromachining related scientific works, due to the fact that lasers have significantly facilitated the production and optimization of small-scale parts and tools. Multiple operations have been successfully realized through laser micromachining, even by using simple equipment and applying plain manufacturing methods, thus proving that the constantly evolving technology may lead to nanometer precision in micromachining processes.

Short-pulsed lasers in particular, combined with the laser ablation process undertaken with their aid, are increasingly used in a broad range of industrial and everyday applications, which include advanced materials processing, large- and small-scale cutting and drilling, micro- and nano-surface configuration, nanotechnology applications, pulsed laser deposition (PLD) of thin films and coatings, surgical operations, and artwork renovation (Balandin, Niedrig, & Bostanjoglo, 1995). In certain situations, the laser–material interaction principles play a significant role in such manufacturing applications as alloy production, soldering, and drilling, as well as in the acquisition of thermo-physical data in high temperatures.

Moreover, widespread attention is nowadays drawn to ultrashort-pulsed lasers, whose pulse duration ranges between a few femtoseconds to picoseconds. Their important advantages over short-pulsed lasers have been frequently exploited by scientists and engineers. These advantages are the minimization of the molten material, the heat transfer and the material's heat-affected zone (HAZ) during and after the ablation process, the efficient energy deposition across the material under radiation, and the high spatial concentration of heat, resulting in higher manufacturing precision.

The laser ablation process has been thoroughly examined both from its experimental and theoretical viewpoints. In the former case, for instance, Pronko, Dutta, Du, and Singh (1995) determined the numerical values of the threshold fluence (energy density) of ablation in gold, in relation to the pulse range between 10 ns and 100 fs. Simon and Ihlemann (1996) studied the ablation of micro- and nano-structures in copper and silica specimens by using short ultraviolet laser pulses, at 248 nm wavelength with a duration between 0.5 and 50 ps. Furthermore, Shirk and Molian (2001) studied the ultrashort-pulsed laser ablation of highly oriented pyrolytic graphite.

Materials Forming and Machining. http://dx.doi.org/10.1016/B978-0-85709-483-4.00006-5

In terms of the theoretical approaches on pulsed laser ablation, numerous efficient methods and computational models have been developed. To begin with, Conde et al. (2004) realized a numerical simulation of the initial laser ablation stages, while simultaneously considering the role of the initial surface roughness of the target in the deviation of the plasma plume. Li et al. (2006) predicted the absorption coefficient and the absorptivity of the target in relation to its temperature and depicted the influence of their dynamic nature in the ablation process. The study of Oliveira and Vilar (2007) employed two-dimensional finite element analysis (FEA) in order to determine the probability of explosive boiling initiation during the laser ablation process on TiC with Nd:YAG and KrF lasers. Fang et al. (2008) presented a thermal model that describes laser ablation on a metallic target, where ultraviolet high-power nanosecond pulses are produced by the laser system. This model is particularly important because it takes the plasma shielding effect into account. Bulgakov and Bulgakova (1999) constructed a thermal model of laser ablation, which draws attention to the formation and heating of the radiation absorbed by the plasma plume, in an effort to calculate the laser energy equilibrium for the duration of a single pulse. In addition, Amoruso, Armenante, Berardi, Bruzzese, and Spinelli (1997) investigated the absorption and saturation mechanisms in plasmas created through laser ablation on Al surfaces. Finally, another finite element modeling procedure conducted by Vasantgadkar, Bhandarkar, and Joshi (2010) showed that various parameters generally affect the laser ablation procedure, such as the laser beam itself—namely the wavelength, fluence, pulse repetition rate, and pulse duration—the material—namely absorptivity, thermal diffusivity, and behavior in melting and boiling—and the layer structure of the specimen on which ablation will be performed.

Through the above, it is evident that the study of various aspects concerning the fundamental mechanisms of laser ablation is still in progress, without any definite conclusions having been drawn as far as the steps, characteristics, and parameters of the process are concerned. Nevertheless, several researchers have incorporated a substantial number of relevant features in their models, aiming to simulate actual situations as reliably as possible.

Recently, molecular simulation models have been significantly developed to cater to researchers' needs in enriching their knowledge on laser ablation details and particularities. These include molecular mechanics (MM), Monte-Carlo (MC) simulation, and molecular dynamics (MD). The latter of these three methods is characterized as deterministic, as opposed to the other two stochastic methods, and its broad use has been encouraged due to its ability to simulate the actual behavior mechanisms of materials, as a result of interactions in the fundamental molecular level. Its ability to predict the temporal evolution of a system comprising interacting particles, e.g., atoms, molecules, or grains, combined with the ways in which it exploits Newton's second law of mechanics to reveal information about dynamic attribute changes in molecules, has made MD simulation a priceless and predominant tool in the hands of any engineer.

The implementation of MD has recently been made possible via various programming environments operating on computers, such as the Java platform, deployed by, among others, Stavropoulos, Stournaras, Salonitis, and Chryssolouris (2010) in order to determine how notable laser ablation characteristics evolve over time, after the incidence of femtosecond beams on iron surfaces. It is worth noting that in this work the possible impact of plasma shielding in the intensity and final results of the process is neglected.

Throughout the present chapter, an MD code is going to be developed within the MATLAB® software package, in an effort to adequately simulate the subsidiary phenomena that occur directly from the moment when a single laser pulse contacts the surface of the target. The results obtained via the theoretical analysis under discussion are compared with experimental results from the application of nanosecond laser pulses on separate spots of thin films made of molybdenum (Mo) and (Al). Configurations of the code are based on the fact that the laser ablation process is undertaken using short pulses, as well as the attenuation of the phenomenon over time because of the expanding plasma plume and the gradually attenuating radiation thereby caused. The objective behind the realization of this analysis is to investigate, at regular time intervals of a few picoseconds, the evolution of such parameters as the number of absorbed photons by the material, the rate of atom removal from the main material (i.e., the ablation rate), the maximum depth of the crater formed due to the application of the beam (i.e., the ablation depth), and the mean temperature of the target under its heating by the beam. The study aims to serve as a strong reference for the researching community of the future, in the attempt to determine the strengths and weaknesses of the simulation and provide possible solutions for the improvement of subsequent MD analyses.

6.2   Basic ablation theory

The term laser ablation is employed to describe the procedure of material removal from a solid or liquid surface, after being radiated by a laser beam. In other words, a small pulsed laser can be focused on the material in order to sufficiently heat it and lead to its vaporization. Pulsed lasers are mostly suitable for this process, although continuous laser beams may also be employed, provided that their intensity is adequately high. The depth at which energy absorption takes place and, as a consequence, the quantity of the ablated material per incident pulse, depend on the optical properties of the material, laser wavelength, and its pulse duration. Specifically, ultrashort laser pulses, normally lasting no more than some picoseconds, are capable of removing material so rapidly, that the material surrounding the created hole absorbs low amounts of heat, making laser-assisted manufacturing processes, such as drilling, feasible, and successful even for materials sensitive to thermal influences.

According to the study of Bulgakov and Bulgakova (1999) on medium fluence lasers (between 1 and 10 J/cm^2), their influence on the material leads to the formation of low-temperature plasma of a 10^4 K order of magnitude, inside which a relatively large part of the laser energy can be absorbed. This affects both the efficiency and the quality of the ablation process, combined with the characteristics of the created plasma plume. As the parameters determining the optical thickness of the plume, such as the electron concentration, the temperature, and the area of the plasma cloud, significantly vary during the whole course of the laser pulse emission, it is necessary to develop thorough computational models in order to achieve a precise description of the radiation absorption potential.

It is difficult to incorporate all procedures that take place during laser ablation within a single computational model. However, it is not particularly difficult to draw general macroscopic observations that may give an initial and general view on the evolution of the ablation process. Initially, laser energy is absorbed by the target, with this energy subsequently transforming itself to thermal vibrations. In case the atoms of the target material have acquired a considerable amount of energy, it is possible that various particles, for example electrons, ions, neutral atoms, or molecules, are emitted on its surface. Moreover, certain circumstances favor the formation of a plasma cloud in front of the target. Laser wavelength and pulse duration, along with the nature of the material, are critical in determining the succession and time scale of the events that occur during the ablation process.

Vasantgadkar et al. (2010) portray the evolution of the laser ablation process, with a plain graphical representation appearing in Figure 6.1. At first, the laser energy is instantly absorbed by the target, leading to its heating. Next, heat transfer occurs within the material through conduction. These two steps constitute phase I, while phase II commences when the surface of the target reaches its melting point, resulting in the fusion, vaporization, and expulsion of the molten material. In general, the rate of heating and the surface temperature are defined through the coefficients of absorption and reflection, and the material's thermal conductivity and specific heat. It is worth mentioning that plasma is released during the final two stages presented in Figure 6.1.

In the situation of picosecond and femtosecond laser pulses, the classical description of laser–material interactions, directly affecting laser ablation analyses, seems to lose their validity because of the nonlinear multiphoton procedures initiated by the elevated intensities of ultrashort pulses, causing absorption levels to rise. Moreover, the respective time scales do not encourage immediate energy transfer from the electron gas to the ion lattice. The above create the necessity for computational models that include, but are not limited to, the two-temperature model (TTM), which consists of thermal descriptions that separate the temperature of the electron from the one of the ion lattice.

Figure 6.2, presented in the analysis of Leitz et al. (2011), demonstrates the differences in the laser ablation mechanism when diverse pulse durations are employed. The first instance depicts the phenomena that take place when short-pulsed laser beams contact the target surface. In the second case, referring to ultrashort laser pulse–material interaction, the minimal reaction time does not allow the material to

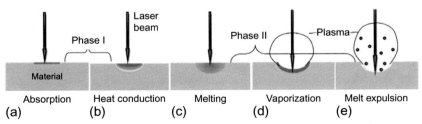

Figure 6.1 Stages of laser–material interaction during pulsed laser ablation (Vasantgadkar et al., 2010).

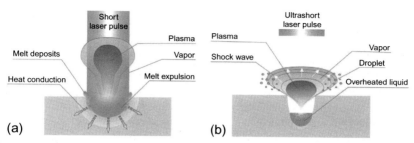

Figure 6.2 Laser beam–material interaction for (a) short pulses and (b) ultrashort pulses (Leitz, Redlingshöfer, Reg, Otto, & Schmidt, 2011).

evaporate continuously, leading to its transition to an overheated liquid state. The result of the above is the appearance of a mechanism named phase explosion, which is characterized by the formation of a mixture of high pressure liquid droplets and steam, undergoing swift expansion.

Most manufacturing processes realized through laser ablation include not only the fusion of the target surface's part that interacts with the beam but also its solidification (Zhigilei et al., 2010). Among the most significant quantities worth studying during laser ablation process investigation is the threshold fluence F_{th}, which is the critical energy density value (in J/cm²) under which laser–material interaction does not lead to the target's ablation. The phenomenon being discussed is generally characterized by the ablation depth caused to the target by each separate incident pulse, something that directly affects the ablation rate quantity (Bozsóki, Balogh, & Gordon, 2011). It has been mentioned by Gordon, Balogh, and Sinkovics (2007) that the threshold fluence, the actual fluence, F, of the incident beam, and the ablation depth, d, relate to each other according to the following equation of logarithmic proportionality:

$$d \propto \ln\left(\frac{F}{F_{th}}\right) \tag{6.1}$$

However, the above equation becomes no longer valid when the energy density values become much higher. In fact, it is replaced by the linear proportionality relation:

$$d \propto \left(F - F_0\right) \tag{6.2}$$

where F_0 is a characteristic value, usually higher than the threshold fluence.

In general, it can be observed that the material removal depth per pulse increases intensely along with laser energy density. However, the possible format and evolution of a biaxial $d(F)$ graph will vary significantly depending on the material, the laser beam wavelength, and, in certain circumstances, the number of incident pulses. Furthermore, the already described approximation is not equally effective for thin metallic films, whose thickness is normally equal to or less than some hundreds of nanometers. The reason for this is that a single incident laser pulse, even of a low energy, is strong enough to fully ablate and degenerate the coating layer from its substrate.

In this case, the diameter of the forming crater on the target material may determine the ablation efficiency in a more suitable manner.

The diverse variables used in an ordinary thermal model of laser ablation can be controlled by either the pulse's energy density or its energy itself. The former case is consistent with excimer laser devices, which produce homogeneous top-hat conic beam profiles (Gordon et al., 2007). On the contrary, the latter of the two parameters best interprets the situation consistent with Nd:YAG lasers, due to the Gaussian spatial distribution of the emitted beam's rays, resulting in actual fluence changes across the beam's cross-section.

As heat transfer inside the target plays a notable role in affecting the results of laser ablation, various relevant equations have been introduced and applied to recent thermal models in order to more accurately simulate the process. A conventional energy balance equation for solid bodies, depending on the mass density ρ, the specific heat $c(T)$ influenced by the temperature T, the thermal heat flux vector $\vec{q}$, and the heat source $Q(\vec{x},t)$ can be used:

$$c(T)\rho\frac{\partial T}{\partial t}+\nabla\cdot\vec{q}=Q(\vec{x},t) \tag{6.3}$$

The value to the right of the equality sign is given by the following relation:

$$Q(\vec{x},t)=-\nabla\left(\vec{S}(\vec{x},t)\right)+U(\vec{x},t) \tag{6.4}$$

where $U(\vec{x},t)$ is the energy provided or needed for the accomplishment of auxiliary chemical reactions, other than the ones directly related to ablation, or phase changes on the material, and $\vec{S}(\vec{x},t)$ is the time average of the Poynting vector.

Through the diverse experimental studies conducted in recent years, a series of common important characteristics of laser ablation has been observed. More specifically:

- When the ablation process takes place under an air environment, a shock wave is produced.
- Explosive ablation leads to various photoproducts, whose nature and quantity is directly dependent on the target material, the wavelength, and the intensity of the laser beam.
- Ordinary laser ablation experiments on metallic targets normally lead to target temperatures around 4000 K, which increase with the radiation intensity. However, this statement is not always valid for nonmetallic targets.
- Laser ablation products typically displace themselves at velocities between 1 and 2 km/s, in a direction perpendicular to the surface of the radiated substrate. The velocity distribution is almost independent of the incident beam intensity.
- The rapid outset and evolution of the material expulsion process results in the formation of an expanding gaseous volume consisting of ablation products, for the duration of the laser pulse.

The material expulsion process may reliably be explained via four separate mechanisms, analyzed in detail by Miller (1994) and Boyd (1992). In brief, the mechanisms studied and their basic traits are as follows:

- *Photothermal mechanism*: It is characterized by the ultrafast transformation of the initial electron excitation of the substrate's particles to heat, leading to the target's temperature increase. More specifically, the de-excitation of the particles is observed and their energy is statistically distributed to the diverse degrees of freedom of the radiated area's particles. There is no definite threshold fluence value associated with the mechanism, because the maximum temperature attained is a linear function of the energy density F. Equation (6.5), where α is the particles' absorption coefficient, F is the energy density of the laser, ρ is the mass density of the material, C_p is the heat capacity per volume unit, and z is the optical penetration depth, representing the temperature increase on the target, within the framework of the photothermal mechanism, justifies the above comments:

$$\Delta T = \frac{a \cdot F}{\rho \cdot C_p} e^{-az} \tag{6.5}$$

- *Photochemical mechanism*: It admits that laser ablation results from the disintegration of many chemical bonds during the process in discussion. The breaking of the bonds causes the photon energy, which exceeds the cohesion energy related to each bond, to be distributed to the separate fragments (atoms) under the forms of kinetic and internal energy. The fragments obtain very high kinetic energies, causing the material's transition to the gaseous phase. Let N_b be the number of disintegrated molecules and φ the quantic efficiency of this disintegration. For a specific beam intensity value I_0, not always equal to the incident beam's intensity, I, the following formula (based on Beer's law) stands:

$$N_b = \varphi\left(I_0 - I\right) = \varphi I_0 \left(1 - e^{ax}\right) \tag{6.6}$$

When I_0 exceeds a certain value, an excessively high number of fragments appear. The forces they exert are high enough to overcome the cohesion forces between the molecules of the material, consequently leading to its ejection. According to the above, the depth of the ejected material, δ, may be approximated by the following formula:

$$\delta = \alpha^{-1} \ln\left(\frac{F}{F_{th}}\right) \tag{6.7}$$

where α is the particles' absorption coefficient.
- *Photomechanical mechanism*: It relates to the instantaneous expansion of the area influenced by laser radiation, because of the intense and rapid increase in temperature caused. The surrounding material does not have the time to adapt itself to this change in volume, something that leads to the development of strong repulsive forces in the heat-affected zone. As a consequence, the material is ejected before it melts. The magnitude of these stresses and forces is influenced by the relation between the energy deposition rate and the characteristic time of mechanical equilibrium in the absorbed volume. Generally, when the laser pulse has a comparable or smaller value than the time required for thermal equilibrium attainment, the heating of the target does not affect its volume, something that results in the development of high thermoelastic pressure. The above is briefly defined by the term inertial or stress confinement.
- *Explosive boiling*: This mechanism appears when the heating rate of the target material is relatively high. It has been interpreted by various researchers as an off-equilibrium phase change. In this case, a criterion inserted in the ablation analysis may determine that either thermal desorption or ablation is responsible for the ejection of the material. In other words, this criterion is employed to compare the formation rate of the uniform bubbles with the evaporative cooling rate. In the case of laser ablation, where the common heating rate values are high, the time

required for the formation and diffusion of bubbles is considerably higher than in the case of lower heating rates. In this case, the temperature rises very fast, up to the point of greatly surpassing the material's boiling point. Therefore, the procedure of regular surface evaporation is transformed into a condition of phase explosion, generally representative of high energy density values. This results in a spontaneous separation of the ejected plasma plume emitted during the laser ablation process, creating a two-phase system that consists of gas molecules and liquid droplets. It has been observed that incident laser energy densities between 40 and 50 mJ/cm² melt the surface of the target, but at least 100 mJ/cm² are necessary to instigate material expulsion. To summarize, explosive boiling takes place when a liquid is overheated in temperatures where the saturation pressure is larger than the one actually exerted on the liquid.

6.3   Plasma shielding theory

Plasma shielding effect plays a predominant role in the analysis of the laser ablation process on metal targets. Plasma is formed because of the sublimation of the target material, which is heated by the incident laser radiation. During the emission of the first and any subsequent pulses, the plasma is further heated and expanded, leading to the creation of a protective shield for the material's upper surface. Due to the plasma plume's influence, the incident laser energy gradually decreases while the plume keeps expanding. Hence, the heating rate of the target decreases, having a similar impact on the ablation rate, the efficiency, and the quality of the ablation process.

Bulgakov and Bulgakova (1999) and Liu and Zhang (2008) have thoroughly studied the plasma shielding phenomenon, emphasizing the case of nanosecond laser pulses. To start with, it is stated by Bulgakov and Bulgakova that intermediate energy densities between 1 and 10 J/cm² not only cause the local vaporization of the material, but also form a plasma plume whose temperature is in the 10^4 K order of magnitude. It is worth noting that the optical depth of the plasma is governed by such parameters as the electron density, the temperature, and the size of the plasma cloud. These parameters significantly vary over time, but possible attempts to model the evolution of the plasma's optical depth demonstrate an extremely high level of complexity if they are to take the aforementioned time dependencies into consideration.

The aforementioned model aims to estimate the energy absorption in the heated plasma and can serve as a basis on which plasma shielding and its consequences can be quantified. According to this model, the density of the absorbed radiation, E_a, is a parameter that can exclusively characterize the absorption increase caused by the heating of the plasma. As the characteristic time of energy exchange between electrons and ions, for temperatures and ion densities of the respective 10^4 K and 10^{19} cm^{-3} orders of magnitude, is much lower than the pulse durations of the nanosecond lasers, it is assumed that the E_a radiation density is absorbed instantaneously and uniformly among all evaporated particles. The absorption of the laser radiation by the target leads to a temperature increase, ΔT, described by equation:

$$\Delta T = \frac{(\gamma - 1) E_a}{kN} = \frac{(\gamma - 1) m E_a}{k \cdot \Delta z \cdot \rho} \tag{6.8}$$

where γ is the effective adiabatic coefficient, k represents the Boltzmann constant, N is the number of evaporated particles (atoms) from a uniform surface, m is the mass of a single atom, and ρ is its mass density.

A formula employed by Bulgakov and Bulgakova (1999) introduces the optical depth of the plasma plume, $\Lambda(t)$, in order to determine the relation between the intensities of the incident laser beam, I, and the beam originally emitted by the laser device, I_0. According to this study, the numerical integration of the absorption, a, across the depth of the target, z, is used to calculate the plume's optical depth. In mathematical terms, Equation (6.9) is employed:

$$I(t) = I_0(t)\exp\left[-\Lambda(t)\right] = I_0(t)\exp\left[-\int_0^{\infty} a(n,T)\,dz\right] \tag{6.9}$$

It is possible to approximate quantity a with a linear product of the particle density, n, and a function $f(T)$ whose value increases with the temperature. This approximation is valid only when the temperature of the plasma is low and the plasma lies in a state of equilibrium:

$$a = n \cdot f(T) \tag{6.10}$$

The last equation can be developed to exclusively contain temperature dependent and first order time-dependent terms. The optical depth of the plume formed is eventually given by the equation below:

$$\Lambda(t) = A \cdot \Delta z(t) + B \cdot E_a(t) \tag{6.11}$$

Quantity $\Delta z(t)$ represents the penetration depth per each applied pulse. The coefficients A and B are governed by the evaporation temperature of the target's atoms, T_v, and are expressed as follows:

$$A = \frac{\rho f(T_v)}{m}, \quad B = (\gamma - 1)\frac{\partial f}{\partial T}\bigg|_{T_v} \cdot k^{-1} \tag{6.12}$$

Figure 6.3 shows the evolution rate and maximum value of the target's temperature, affected by the fluence of the laser beam, both by considering and ignoring the plasma shielding effect for various laser fluences. If this phenomenon is taken into account, then the peak absolute temperature of the target generally drops 4–6 times, compared to the case where this effect is not considered.

The temperature evolution due to the plasma shielding effect can be explained as follows: the temperature increase on the target causes equivalent changes to the number of vaporized particles on the material. This subsequently augments the energy loss on the target surface. Moreover, the energy absorbed by the plasma gradually increases with temperature; in this case, the effective laser fluence of the incident beam decreases. That is consequently reflected on the temperature of the target, which also

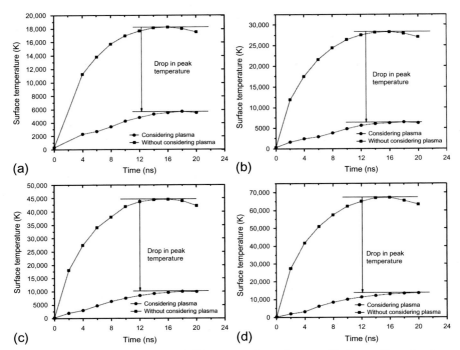

Figure 6.3 Temperature change on a TiC target surface, when considering and ignoring the plasma shielding effect (Vasantgadkar et al., 2010).

diminishes. For instance, by observing the (c) and (d) graphs of Figure 6.3, it is apparent that the peak value of the target's temperature rises by 1.5 times (from 40,000 to 60,000 K) when laser fluence increases from 10 to 15 J/cm² , thus satisfying a relationship of direct proportionality. Conversely, a similar fluence rise leads to a significantly less intense increase in temperature, with respect to both its general evolution and its peak value (from 8679 to 10,520 K).

6.4   MD simulation in laser ablation

The MD simulation method was conceived and developed in 1956 by Adler and Wainwright (1957) because of the need to include the time variable in the evolution of a system under investigation. Unlike the MC and MM methods, which are regarded as stochastic, MD is a deterministic method through which the actual behavior mechanisms of materials are simulated, taking into account the interparticle forces in the fundamental molecular field. Its use in laser applications is extensive, especially when it comes to target ablation by short-pulsed (Gamaly et al., 2005) and ultrashort-pulsed lasers (Bulgakova et al., 2007; Ohmura, Fukumoto, & Miyamoto, 2000). MD enables the calculation of the equilibrium and transfer properties in a many-body classical system. The system's energy, temperature, and pressure fall into the first category, whereas the second category comprises the diffusion coefficient, the shear viscosity, and the thermal conductivity of the system.

The MD method may frequently be based on the following concept: from the moment that any given material or substance is formed by fundamental particles, it is possible to use statistical methods in order to establish the macroscopic natural properties of the materials or substances, as long as the basic dynamic parameters of their component particles have already been determined. The properties may be found through the averaging of all measurements conducted within an MD simulation. Additionally, all transfer properties can be calculated through the acquired data, because the full trajectories of the particles are already at the researcher's disposal. These properties are defined as time-dependent correlation functions, in the atomic level. In general, a typical MD simulation for a many-body system can take place in three discrete steps:

- Determination of the initial positions and moments of the system's particles. These interact with each other under an embodied potential. Therefore, this potential defines the extent to which the simulation results represent the system of interest.
- Advancement of the system via Newton's second law of mechanics, whose general format presents the force exerted on each particle i as being directly proportional to its acceleration and mass, according to the formula $F_i = m_i \cdot a_i$.
- Calculation of natural quantities as a function of the particles' positions and moments developed over time.

Most MD simulations overall require identical courses of action to be followed, in comparison with related experimental procedures. Firstly, the specimen is prepared through the optimum and most reliable choice of the physical system under description. As the MD method is based on the solution of Newton's equations, iterative solution of these equations will be required in a system of N particles (i.e., atoms) until the point where no further changes in the system's properties occur over time. Finally, the desired measurements are taken after the system has fallen into a state of total equilibrium.

The calculation of the interatomic forces during an MD simulation is made possible through an interparticle interaction potential function, $U\left(\vec{r_1}, \vec{r_2}, \cdots \vec{r_N}\right)$, which depends on the displacement $\vec{r}$ of each of the system's N particles. In the majority of situations, potential U must serve as the exclusive input data for the calculation of the desired quantities during the simulation. Especially as far as the forces are concerned, it is only necessary to compute the gradient of the potential, as the following relation suggests:

$$\vec{F_i} = -\vec{\nabla_i} U\left(\vec{r_1}, \vec{r_2}, \ldots \vec{r_N}\right)$$
(6.13)

Although the MD simulation procedure is advantageous in many aspects, it contains two very important limitations. The first one is the need to develop a mathematical formula that describes the interaction potential with maximum accuracy. This has been partially resolved by introducing various empirical formulas, such as the Stillinger and Weber (1985) potential. The second one deals with the maximum number of atoms or molecules that a system can accommodate for the successful completion of the simulation process. Although 10–100 million atoms can be inserted into a system without negatively affecting the simulation results, this number is greatly affected by the required simulation time. For example, the need to acquire statistical information about the phonons, which is a procedure that involves large amounts of time, signifi-cantly increases the simulation times in large-scale systems, such as one that contains 5×10^7 Si atoms organized in a cubic simulation volume with an edge of 1000 Å.

Recently accomplished researches and numerical simulations have employed var-ious forms of the MD method, which can be classified in the two broad categories of classical and hybrid MD. The former category uses equations of classical mechanics in order to describe and examine the atomic and molecular systems and interactions. The latter have been introduced in an effort not only to extend the simulation volume further than in the case of classical MD, but also to optimize the observation of the processes that take place during laser ablation, both macroscopically and microscop-ically (e.g., in the nanoscale). The accomplishment of a hybrid MD simulation is re-alized via the connection of a typical MD simulation volume with another one using either FEA or the TTM, leading to the widely used MD-FEA and TTM-MD methods, respectively.

Most classical MD descriptions employ either the embedded-atom method (EAM) or the pair potential approximation (PPA). The EAM approximation has been used by Nedialkov and Atanasov (2006) in investigating the ablation of Fe targets due to the incidence of femtosecond laser pulses, combined with the laser-induced deep drilling process. This semiempirical expression helps identify the factors that influence the final geometrical shape of the hole formed during laser ablation. The first one is the deposition of the ablated material at a specific height of the hole's internal borders, where the hole becomes narrower. The second one refers to the further degeneration of particles that interact with the ones originally removed, referred as secondary ab-lation. EAM was used by, among others, Schäfer, Urbassek, and Zhigileiei (2002) in an attempt to simulate the ablation of Cu targets using picosecond laser pulses, and by Yamashita, Yokomine, Ebara, and Shimizu (2006), who investigated the heat transfer and shock wave formation processes during femtosecond laser ablation on a thick Al target, using 80,000 Al atoms for the MD simulation conducted.

PPA requires much less time than EAM, because possible simultaneous interac-tions between three or more atoms are ignored, something that reduces calculation times without impeding the engineer's attempts to accurately and precisely describe the properties of the material under investigation. The 12-6 Lennard-Jones expression is a highly representative example of the PPA application, calculating the potential using a formula that depends on the material's stress and strain parameters, in other words, σ and ε, respectively, as well as on the distance r between two adjacent atoms. It is an adaptation of the Mie potential relation, for $n = 12$ and $m = 6$:

$$u(r) = \frac{\lambda_n}{r^n} - \frac{\lambda_m}{r^m}\bigg|_{m=6}^{n=12} = 4\varepsilon\left[\left(\frac{\sigma}{r}\right)^{12} - \left(\frac{\sigma}{r}\right)^6\right] \tag{6.14}$$

The alternative of using the Morse potential function (MPF) has been employed in a wide range of applications, including the interaction analysis between laser beams and materials of a metallic and organic nature. The Morse potential is expressed through the following formula:

$$\varphi\left(r_{ij}\right) = D\left[e^{-2a\left(r_{ij}-r_0\right)} - 2e^{-a\left(r_{ij}-r_0\right)}\right] \qquad (6.15)$$

where r_{ij} is the distance between two adjacent atoms, r_0 is the equilibrium distance, D is the dissociation energy, which is the minimum energy value required for the breakup of a bond between two atoms, and a is the dissociation constant.

The first derivative of Equation (6.14) in terms of the distance r gives the interaction force between particles, F. The above formula has been constructed in such a way that the force is attractive ($F < 0$) for high distances, repulsive ($F > 0$) for low ones and nil ($F = 0$) for a given equilibrium distance r_0. For this distance, the potential force $P(r)$ must acquire its minimum value.

With concern for researchers' studies on laser ablation aspects using the MPF, Cheng and Xu (2005) investigated the disintegration mechanisms of Ni crystals during laser ablation, comparing and contrasting their differences according to the value of the laser energy. Zhigilei, Kodali, and Garrison (1997) presented a breathing sphere model, used in conjunction with the MPF method in order to study the ablation of organic molecules. Nedialkov, Imamova, Atanasov, Berger, and Dausinger (2005) developed a theoretical MD model and applied it on three different materials, namely, Al, Ni, and Fe, in an attempt to determine how the laser ablation mechanisms differ with pulse duration (i.e., 0.1, 0.5, and 5 ps), wavelength, 248 or 800 nm, and fluence from the threshold fluence until 0.5 J/cm². The ablation depth attained during laser radiation of the target is generally much higher in the Ni specimen. Moreover, the MD simulation results are in almost total accordance with the experimental findings for the Al specimen, but this does not always happen with the Ni target, with the experimental ablation depth values being higher when laser energy density rises above 0.1 J/cm².

When MD calculations are performed under such circumstances that all the atoms and molecules always remain in the solid phase, there are no constraints with concern to the simulation volume. However, this is not the case when liquid or gaseous parts interfere in this volume. If the simulation volume is not constrained, a thorough analysis on each of the system's dimensions will be required, something that has a highly negative impact on the calculation time. The above problems are resolved by introducing boundary conditions in the MD system. This aids in simulating the properties of materials by considering a reference computational cell as a part of the infinite crystal and formulating a series of equations that express the characteristics and properties of the particles lying outside the cell. In brief, the three different boundary condition types encountered are the following:

- *Free boundary conditions (FBC)*: When these conditions are applied, they allow any particle in the simulation volume, which will come in contact with the theoretical surface determined by these conditions, to be removed from the simulation volume. The same conditions enable any particle initially not belonging to the volume, to eventually enter it.
- *Reflective boundary conditions (RBC)*: These are characterized by a theoretical surface which, when touched by a particle of the simulation volume, reverses the perpendicular to

the surface vector of the particle's velocity. This incident happens because of the oscillating movement of the crystalline structure's atoms.

- *Periodic boundary conditions (PBC)*: These conditions let the infinite crystal constitute a periodic repetition of the reference computational cell toward all dimensions of the three-dimensional space (i.e., across the *X*, *Y*, and *Z* axes). Consequently, it is assumed that the crystal comprises a countless number of particles, each playing the role of a reflection for the reference computational cell's particles. In other words, the insertion of PBC actually serves as the creation of identical replicas of the reference computational cell, used for undertaking the MD simulation around it. For the duration of the simulation, the vibration of the solid's crystalline structure will oblige some of the particles to exit the simulation volume by following the appropriate trajectories. However, the compulsory requirement to preserve the number of particles inside a cell leads to the need to re-introduce a new particle inside the simulation volume. The new particle must be characterized by the same velocity and potential energy as the one that has just been removed from the cell.

6.5 Description of the modeling process

A theoretical MD model capable of reliably simulating the laser ablation process of metals generally needs to fulfill two principal requirements. First of all, it is necessary to describe both the position of each atom and the positions in which two adjacent atoms interact with each other. Secondly, the energies of the atoms must be calculated and the boundary conditions that accompany the problem have to be determined. Next, all ablation stages up to the attainment of thermodynamic equilibrium, under environmental temperature, pressure, and humidity, have to be simulated.

All parameters utilized within the current analysis underwent a process of nondimensionalization. The reason for this approach is the considerable difference in orders of magnitude between measurement units employed both in the macroscopic and the atomic levels. Therefore, it is necessary to avoid possible confusion between measurement units belonging to different systems and to enclose all numerical values in such a range that does not risk leading to potential problems in precision of calculations. Nondimensionalization in the current MD modeling process is undertaken in six fundamental quantities: length, time, mass, energy, velocity, and temperature; all other units are secondary and derive from the aforementioned primary ones.

For the potential energy of a pair of particles, the MPF is given by Formula (6.14), provided that their distance is less than a predetermined cut-off radius, r_c. For higher distances, the potential energy, $P(r_{ij})$, between the adjacent atoms i and j equals zero. The kinetic energy of each atom is obviously calculated by multiplying its mass with the squared value of its vector speed and dividing the product by 2. Hence, the total energy is a result of the kinetic and potential energies' addition. For each pair of i, j atoms, the following equation is in operation:

$$E_{\text{tot},i} = D \sum_{\substack{j=1 \\ j \neq i}}^{N} \left[\exp\left[-2a\left(r_{ij} - r_0\right)\right] - 2\exp\left[-a\left(r_{ij} - r_0\right)\right]\right] + \frac{1}{2} m_i v_i^2 \tag{6.16}$$

However, the Morse coefficient D and the mass- and velocity-dependent term may disappear during nondimensionalization, thus facilitating the MD modeling process in general.

PBC are applied on the XY plane, across which the target material is supposed to extend. FBC are applied on the upper surface of the material in order to allow the removal of individual particles or atom groups from the target. Finally, RBC on the lower surface ensure that the energy, volume, and total number of particles within the atom system are preserved throughout the entire simulation.

Solving the Newton equation that relates to the movement of the target's atoms is a procedure that initially requires the atom system to be equilibrated. During equilibration, the vector velocities of all atoms are chosen in such a way that the total momentum of the system is equal to zero, and that the Maxwell–Boltzmann (MB) velocity distribution is satisfied (Chiang, Chou, Wu, & Yuan, 2006). For reasons of simplicity, the Gauss distribution of the velocities is going to be employed, given that its format and shape, in the atom amount–velocity domain, is not very different from the one of the MB distribution. Moreover, it has been tested in MATLAB® that the usage of the Gaussian distribution aids in attaining the desired system equilibrium under a standard environmental temperature value, such as 300 K.

The above testing was realized by introducing a velocity convergence function, conceived by Schommers (1986). For a system of N particles, with three-dimensional vector velocities $\overrightarrow{\upsilon_t}$ defined by their x, y, and z constituents, the convergence function $C(t)$ takes the following form:

$$C(t) = \frac{\frac{1}{N}\sum_{i=1}^{N}\left[\upsilon_{ix}^2 + \upsilon_{iy}^2 + \upsilon_{iz}^2\right]^2}{\left[\frac{1}{N}\sum_{i=1}^{N}\left[\upsilon_{ix}^2 + \upsilon_{iy}^2 + \upsilon_{iz}^2\right]\right]^2} \tag{6.17}$$

For the purpose of the current analysis, the criterion $\left|C(t) - \frac{5}{3}\right| \leq 0.25$ is satisfied for the desired thermodynamic equilibrium to be achieved. This leads to a slight precision degradation in the simulation without affecting its quality whatsoever. Incidentally, Schommers (1986) proposes the critical value 0.2 instead of 0.25.

During the course of the MD simulation, it is necessary to make the atom system always achieve equilibrium after the completion of each simulation time step. A classical Verlet algorithm or a predictor–corrector method would be relatively suitable to serve the desired purposes; however, the intended goal may be best fulfilled by employing a Leapfrog Verlet algorithm, due to its good numerical stability, its simplicity, and convenience, as well as its satisfactory computational cost. Let x_n, υ_n, and α_n be matrices that describe the positions, velocities, and accelerations of a particle during the n time moment. For a given time step δt, the Leapfrog Verlet algorithm may be expressed as follows:

$$\upsilon_{n+\frac{1}{2}} = \upsilon_{n-\frac{1}{2}} + \alpha_n \delta t \tag{6.18}$$

$$x_{n+\frac{1}{2}} = x_n + \upsilon_{n+\frac{1}{2}} \delta t \tag{6.19}$$

The above relation can be developed in the three-dimensional domain, which leads to the following:

$$\begin{bmatrix} \upsilon_{ix} \\ \upsilon_{iy} \\ \upsilon_{iz} \end{bmatrix}_{n+\frac{1}{2}} = \begin{bmatrix} \upsilon_{ix} \\ \upsilon_{iy} \\ \upsilon_{iz} \end{bmatrix}_{n-\frac{1}{2}} + \begin{bmatrix} \alpha_{ix} \\ \alpha_{iy} \\ \alpha_{iz} \end{bmatrix}_n \delta t \tag{6.20}$$

$$\begin{bmatrix} x_{ix} \\ x_{iy} \\ x_{iz} \end{bmatrix}_{n+\frac{1}{2}} = \begin{bmatrix} x_{ix} \\ x_{iy} \\ x_{iz} \end{bmatrix}_n + \begin{bmatrix} \upsilon_{ix} \\ \upsilon_{iy} \\ \upsilon_{iz} \end{bmatrix}_{n+\frac{1}{2}} \delta t \tag{6.21}$$

In the Leapfrog Verlet algorithm, positions during the n, $n+1$, $n+2$, etc., time moments are known, something that is not the case for velocities, for which information exists only during the $n-1/2$, $n+1/2$, and $n+3/2$ times. It is possible to perform linear interpolation between the $n-1/2$ and $n+1/2$ time moments by averaging the relevant velocities in order to come up with the υ_n velocity for the desired moment n.

Before the application of the Leapfrog Verlet algorithm, it is vital to computationally model the laser radiation process on the target material, which is assumed to begin after the initial equilibrium condition has been fully accomplished. The procedure of approaching and simulating the radiation is mainly carried out in terms of energy. The beam is simulated both in the time and space domains, following different distributions for each. The space profile of the beam follows the Gaussian distribution, while time profile is uniform. Beam's modeling is complemented via the application of the Beer–Lambert law, which is favored because of the fact that the beam modeling involves both its reflectance, due to the material's optical properties, and laser beam waste index. This index is not going to be considered because the beam's radius does not change over time with regards to its initial value. Ignoring the index is also a result of using dimensions of a nanometer order of magnitude during the present study.

For a $t_p = t_2 - t_1$ laser pulse duration and a $t_0 = \dfrac{t_1 + t_2}{2}$ time center of the pulse, the term that expresses the Gaussian temporal distribution of the beam is expanded as follows:

$$S_{p,\text{Gauss}} = \frac{1}{t_p \sqrt{\pi}} \exp\left(-\left(\frac{t - t_0}{t_p} \right)^2 \right) \tag{6.22}$$

In terms of the beam's spatial distribution across the Z axis, two important factors have to be taken into consideration: the plasma shielding effect, expressed by Equations (6.9) through (6.12), and Beer–Lambert's law widely used to describe the attenuation of the beam's intensity when penetrating into the target material. According to this law, the intensity of the radiation is dependent on the one that the beam has on the target's surface ($z=0$). For a given absorption coefficient β that characterizes the material, the formula that represents the above law is the following:

$$I(z) = I_{z=0} \exp(-\beta z) \tag{6.23}$$

Assuming that the radiated surface has a circular cross-section, $A = \pi (r(z))^2$, and taking into account the fact that the energy of the radiation is a sum of all the energies that the beam's photons entail, $E_v = N_p E_p = N_p hv = N_p h \dfrac{c}{\lambda}$, it is possible to conclude that the number of photons changes with the depth inside the target material according to the formula below:

$$N_p = \frac{P_L \delta t \lambda}{hc} \exp(-\beta z) \tag{6.24}$$

where P_L represents laser power, δt is the time step, λ is the beam's wavelength, h is the Planck coefficient, and c is the speed of light in vacuum, given the fact that the above formula applies to the beam before its penetration inside the material, where the speed of light is lower.

The Gaussian spatial distribution of the beam in the XY plane is attained by using a two-dimensional variable, which depends on the photons' positions on the x and y axis, the mean values μ and standard deviations σ for each of the axes, and the correlation coefficient ρ. The general format of such an equation is presented by Stavropoulos et al. (2010) but may be further simplified according to various assumptions. Firstly, if the geometry of the beam is circular, with a radius symbolized as r_b, it is obvious that the standard deviation is the same, regardless of the axis. Secondly, if the center of the laser beam on the XY plane is considered as a reference point, the mean value μ for the Gaussian distribution in discussion is zero. A precision parameter a_c may be used in order to confront potential precision problems that arise from the use of the Gaussian distribution. When the above parameter is multiplied with the standard deviation value, it gives the diameter of the laser beam. Thus:

$$a_c \sigma = 2r_b \Rightarrow \sigma = \frac{2}{a_c} r_b \tag{6.25}$$

The above statements lead to the simplification of the $f(x,y)$ function, which finally receives the format:

$$f(x,y) = \frac{a_c^2}{8\pi r_b^2} \exp\left[-\frac{a_c^2}{8 r_b^2} (x^2 + y^2) \right] \tag{6.26}$$

The integration of (6.26) across the x and y axes is capable of providing the possibility for a photon to lie in a given (x,y) location. With r_a being the atomic radius, this possibility is initially expressed as follows:

$$\prod (x - r_a \leq X \leq x + r_a, y - r_a \leq Y \leq y + r_a) \tag{6.27}$$

The normalization of (6.27) leads to a relatively complex relation, which is implemented into MATLAB® using the cumulative probability function Φ. The respective line of code will be based on the formula below:

$$\Pi\left(a_c \frac{x-r_a}{2r_b} \leq a_c \frac{X}{2r_b} \leq a_c \frac{x+r_a}{2r_b}, a_c \frac{y-r_a}{2r_b} \leq a_c \frac{Y}{2r_b} \leq a_c \frac{y+r_a}{2r_b}\right)$$
$$= \left(\Phi\left(a_c \frac{x+r_a}{2r_b}\right) - \Phi\left(a_c \frac{x-r_a}{2r_b}\right)\right)\left(\Phi\left(a_c \frac{y+r_a}{2r_b}\right) - \Phi\left(a_c \frac{y-r_a}{2r_b}\right)\right) \tag{6.28}$$

If R is the index of reflectivity, then the one of absorptivity is evidently $1-R$. The general equation that eventually provides the absorbed energy of the material is the following:

$$S = (1-R)\left[P_L \delta t \exp(-\beta z)\right]\left[\frac{a_c^2}{8\pi r_b^2} \exp\left(\frac{a_c^2}{8\pi r_b^2}(x^2 + y^2)\right)\right] \tag{6.29}$$

Moreover, multiplying the number of absorbed photons N_{AP} with the energy of each separate photon, gives the total photon energy absorbed by the particles. This energy is added to the one currently possessed by the particles before the radiation process starts.

In order to determine, during the theoretical analysis, whether an atom will be removed from the simulation volume, the cohesion energy criterion is going to be applied. According to this, if the sum of the atom's kinetic energy and energy absorbed from the photons, distributed according to the previous mentions, is higher than the atom's cohesion energy, C, the atom will be removed. In the inverse situation, the atom will remain in the target material and a new velocity will be calculated for it. Despite the fact that a series of MD studies present the cohesion energy as a value that depends on the number of atoms contained in the nanostructure's crystal, it is currently easier to specify, as a general cohesion energy value, the one that characterizes a given material macroscopically. For Mo, this value equals $-6.82\,\mathrm{eV}$, while the value is considerably different in the case of Al, with a reliable estimate being $-3.45\,\mathrm{eV}$.

The new velocities of each atom are calculated by taking the new kinetic energy that it has acquired after the laser radiation process and solving the standard kinetic energy equation in terms of the velocity, v. The Leapfrog Verlet algorithm can subsequently be employed in order to advance each atom's characteristics by a single time step, which is chosen to be equal to some picoseconds, or small segments of the entire duration of a nanosecond laser pulse.

After the process of recalculating the new positions and velocities of the atoms, the following relation may be used to determine the interparticle forces between two adjacent atoms in each of the three Cartesian axes:

$$\begin{bmatrix} f_{ix} \\ f_{iy} \\ f_{iz} \end{bmatrix} = 2aD \sum_{\substack{i=1 \\ j\neq i}}^{N} \begin{bmatrix} x_j - x_i \\ y_j - y_i \\ z_j - z_i \end{bmatrix} \frac{\left[-\exp(-2a(r_{ij} - r_0)) + \exp(2a(r_{ij} - r_0))\right]}{\sqrt{(x_j - x_i)^2 + (y_j - y_i)^2 + (z_j - z_i)^2}} \tag{6.30}$$

Using the above equation, it is now easy to calculate the desired acceleration values in order to successfully employ the Leapfrog Verlet algorithm. Dividing the forces by the atomic mass can lead to the desired outcome.

To summarize, the computational procedure followed during current MD analysis, which has been implemented in MATLAB®, comprises the following steps:

- System preparation
- System equilibration
- Loading simulation
- Extraction of results

A detailed flowchart that describes the step-by-step process toward the desired solution is presented in Figure 6.4.

In the effort to construct a sturdy MD code able to satisfy the majority of the requirements for a successful theoretical laser ablation simulation, MATLAB® was preferred to other programs, because of its high usefulness in simulating phenomena and finding analytical solutions for a series of theoretical and practical problems, especially those pertaining to nanosystems. The main advantages that favor MATLAB® over other packages include the abundance of commands and functions that can be introduced in the computational model, the high execution speed for even high-length codes, and the possibility of applying various molecular simulation methods, including MD.

6.6   Experimental results

Theoretical analysis on laser ablation, undertaken through the MD simulation described in the previous sections, needs to be carried out in conjunction with the realization of experiments during which the process is performed in real-life situations. In order to accomplish this goal, two different thin films made of Mo and Al on glass substrates were used as target materials on which nanosecond laser pulses were applied, as part of a scribing process leading to local ablation of the material. The experimental setup used is installed in the Laser-based Techniques and Applications Lab (LATA) of the Theoretical and Physical Chemistry Institute of the National Hellenic Research Foundation and is presented in Figure 6.5.

A Quantel Mo, YG851 Nd:YAG laser ($\lambda = 355$ nm, FWHM = 10 ns) was employed. The 355 nm laser beam passes through a beam splitter, used to reflect and refract part of the incoming radiation, thus reducing the beam's energy without altering its shape. After the beam passes through a diaphragm, which decreases its diameter, it enters a focusing lens before its incidence on target material. The specimen is placed on a moving platform, whose movement is controlled through a stepper motor. The LIPS technique was used to optimize the focusing conditions of the laser beam during scribing. Despite the fact that the repetition rate of the laser is set to 10 Hz by default, its reduction to 2 Hz allows for better control of the procedure.

What is intended to be examined during the experiments is the form of the radiated spot on the specimen after a single laser pulse has interacted with the target material.

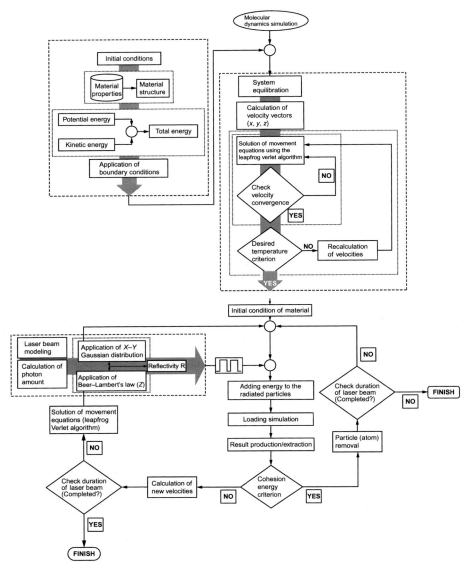

Figure 6.4 Logical flowchart for the development of the MD simulation code, used to model the pulsed laser radiation and ablation processes.

During the experiments, a total of six different spots were chosen on the target per each separate situation, for subsequent irradiation by the laser beam. Each pair of spots was irradiated with a laser beam of a different energy, ranging between 3, 2, and 1 mJ/pulse. These values were proven sufficiently capable to ablate any region on the specimen without having to apply an additional pulse other than the first one. The three cases investigated during the experiment are characterized by differences in

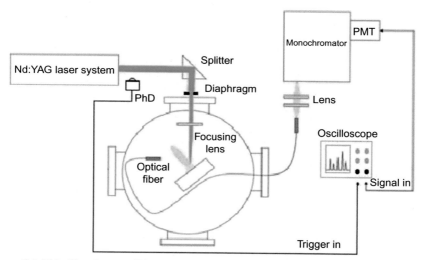

Figure 6.5 Thin films laser scribing setup and LIPS diagnostics.

the irradiated material and the diameter of diaphragm through which the beam passes before hitting the target. These are the following:

- *Case 1*: Mo target, 3 μm diaphragm diameter
- *Case 2*: Mo target, 1.5 μm diaphragm diameter
- *Case 3*: Al target, 1.5 μm diaphragm diameter

Mo and Al specimens have a length of 1.3 mm and width of 0.8 mm, because of the mask that was used during the PLD deposition of the thin films. The Mo film is 280 nm thick and the Al film is 390 nm thick. After completion of the experiments, an optical microscope was employed in order to closely focus on the ablated area and the HAZ that surrounds it. Thanks to this microscope, it is easy to make observations on the dimensions of the ablated area, but this is not the case for the material's depth and, thus, the ablation depth was attained after the relevant process has been finalized. However, a qualitative indication of how deep the beam has penetrated may be given by the color of each region on the target, as viewed through the microscope. Moreover, a KLA-Tencor Alpha-Step IQ Surface Profiler was used in determining the ablation depth, as a result of the ablation process.

6.6.1 Case 1: Mo target, 3μm diaphragm diameter

Figure 6.6 depicts the form of the ablated region after the irradiation of the target with 355 nm laser beam of (a) 3 mJ/pulse, (b) 2 mJ/pulse, and (c) 1 mJ/pulse energies. The dark gray color on each surface is characteristic of the molybdenum material and indicates that the degree of ablation is very high or even total (100%), meaning that the beam has reached the entire depth of the material in each separate position. Generally, the color tint demonstrates the depth to which the beam has penetrated. Areas characterized by an almost white tint, showing discoloration, demonstrate that they have

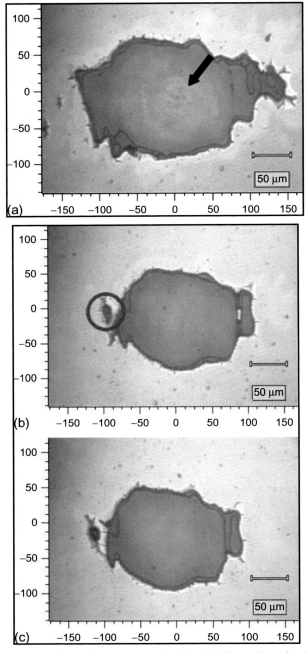

Figure 6.6 Micrographs of the ablated spots on the Mo thin film surface, for diaphragm diameter of 3 μm and incident laser energy values of (a) 3 mJ, (b) 2 mJ, and (c) 1 mJ.

been affected by the laser beam, but not to such an extent that the laser–material inter-action leads to ablation of particles.

The first important observation from Figure 6.6 is the fact that the decrease in laser energy causes the ablated surfaces to become more uniform and smooth in terms of shape and ablation depth. The latter is justified by the darker shade of the material and the less intense differences in the color tint, apparent in Figure 6.6b and c images. Micrograph (a) shows a smear in the center of the ablated surface. This indicates that the amount of laser radiation that affects the respective area is significantly higher than the one affecting peripheral regions of the surface and also suggests that the material has undergone local burning. This is because of the Gaussian profile of the laser beam of the Nd:YAG laser system.

It is also apparent that the breadth of the HAZ decreases with laser energy, be-coming very small in Figure 6.6c, after the irradiation of the target with a 1 mJ/pulse. This denotes that lower laser energy causes rays to be more concentrated around the intended beam application spot, which normally coincides with the center of the beam. From all the above, it is obvious that high laser energies do not have a positive impact on the quality of the ablated area.

6.6.2   Case 2: Mo target, 1.5 μm diaphragm diameter

In Figure 6.7, three different spots from the ablated surface are presented, after its laser irradiation with a beam passed through a diaphragm of half the diameter of that in Case 1.

A first comparison between the matching (a), (b), and (c) images of Figures 6.6 and 6.7 shows that the width of the ablated region becomes significantly smaller after the diaphragm size is halved; this observation also applies to HAZ. The above can be explained by the fact that the reduction of the diaphragm size leads to a higher concentration of the laser radiation around the position where the energy is maximum (i.e., the center of the beam) as the Gaussian spatial profile demands. Mathematically, it is possible to say that this phenomenon results in the creation of a new Gaussian distribution for the beam, in which the mean value practically remains unchanged and the standard deviation decreases. The extensions/bulges are generally less intense in the screenshots of Figure 6.7, something that is a natural consequence of the narrower diaphragm.

The objective of Table 6.1 is to present and compare important results taken from the experimental ablation of all Mo spots, satisfying the conditions set in Cases 1 and 2. The numerical values shown in Table 6.1 refer to the area of the ablated re-gion, laser fluence, HAZ area, ratio between ablated and HAZ areas, and ablation efficiency expressed as the volume of the thin film, assuming a 280 nm thickness, in total compliance with the actual experiment, divided by the laser energy.

In general, a reduction in the laser energy value similarly affects the area of both the ablated region and the HAZ, although the latter is increased when the diaphragm diameter and laser energy receive their minimum values. In fact, very low energy, combined with low diffusion of rays induced because of the small diaphragm, leads to a HAZ that is significantly larger compared to the main ablated region. Conversely,

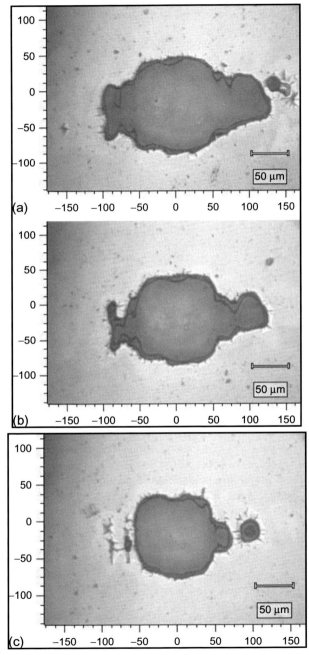

Figure 6.7 Micrographs of the ablated spots on the Mo thin film surface, for diaphragm diameter of 1.5 μm and incident laser energy values of (a) 3 mJ, (b) 2 mJ, and (c) 1 mJ.

Table 6.1 Presentation and comparison of important quantities derived from laser ablation experiments on Mo thin film

Case no.	Laser energy (mJ)	Ablated area (μm²)	Laser fluence (J/cm²)	HAZ area (μm²)	Abl:HAZ area ratio	Ablation efficiency (μm³/mJ)
1A	3	32,642.8	9.19039	19,293.1	1.69194	3046.6
1B	2	20,951.5	9.54586	14,547.4	1.44022	2933.2
1C	1	18,970.5	6.27134	12,335.5	1.53788	5311.7
2A	3	21,859.8	13.7238	11,106.1	1.96827	2040.2
2B	2	21,766.3	9.18852	11,392.0	1.91067	3047.3
2C	1	9937.8	10.0626	13,977.0	0.71101	2782.6

higher laser energy values favor the realization of extensive material ablation with the respective area even reaching values double those of the HAZ, when the diaphragm takes its minimum size. Moreover, a wider dispersion of the rays across the *XY* plane, permitted by the larger diaphragm, increases the area of the HAZ, making its percentage difference with the ablated area smaller.

In addition, the irradiation of the target with the highest possible energy and the least possible ray diffusion (Case 2A), attained through the 1.5 μm diaphragm, ensures that the highest amount of energy interacts with any given area of the target. The quantity that justifies the above is the laser fluence, expressed as energy per surface unit and being equal to over 13 J/cm² for the case in discussion. In Case 1C the lowest fluence value is encountered, slightly over 6 J/cm². By observing the trend of laser fluence and ablation efficiency values in Table 6.1, it can be seen that a rise in fluence deteriorates the process in terms of efficiency.

The Alpha-Step IQ profilometer is able to give valuable information on the maximum ablation depth attained after the end of the scribing process. According to Figure 6.8, it is possible for the Mo thin film to be ablated across its entire thickness, regardless of the diaphragm size and laser energy employed. The curves lying under the zero horizontal line become more uniform as laser energy decreases, thus showing that the surface profile becomes more homogeneous and the fluctuations in the ablation depth gradually smoothen. All measurements pertaining to ablation depth were realized on the spots radiated with laser beams passed through a diaphragm of 1.5 μm diameter.

It is worth noting that the maximum surface roughness measurements showed that the respective quantity increases while the laser beam becomes less intense. This mainly happens because of the extensive formation of HAZ that is not always accompanied by material removal, as well as because of the solid or molten material expulsion throughout the process. The roughness was measured equal to 0.63092 μm for a 3 mJ energy, 1.1229 μm for a 2 mJ energy, and 1.2652 μm when the energy equals 1 mJ.

6.6.3 Case 3: Al target, 1.5 μm diaphragm diameter

The case of the Al specimen exhibits a number of important differences compared to the two previous situations, primarily because of the material's unique nature. The most

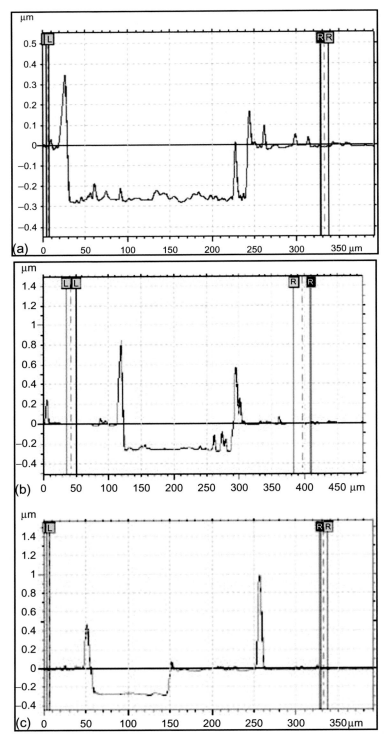

Figure 6.8 Alpha-Step IQ profilometer digraphs for the ablation depth and maximum surface roughness on the Mo thin film for laser energy values of (a) 3 mJ, (b) 2 mJ, and (c) 1 mJ.

serious difference, which is bound to greatly affect both the ablation process and the irradiated region's final form, is the index of reflectance, R. The higher it is, the less laser energy will be absorbed by the specimen, making the process of material removal much more difficult due to the beam's weakness in heating the specimen to such an extent that sufficient material can be later ablated. Bass, Van Stryland, Williams, and Wolfe (1995) present a diagram on a two-dimensional axis system in which the reflectance of three materials, namely Al, Au, and Ag, is presented for different wavelength values between 200 nm and 5 μm. The data of the diagram imply that the relevant reflectance value, R, for Al, for a 355 nm wavelength is equal to 92%. In the case of Mo, the absorptivity is calculated equal to 57.14%.

Figure 6.9 presents the irradiated areas belonging to Al specimens. It is initially obvious that the irradiated area is generally smaller in the Al specimen. The fact that the ablation process has only taken place partially may be indicated through the color tint of the irradiated region, which is considerably lighter on the Al target than on the Mo target. It can be perceived that the relatively low incident laser radiation on the Al target specimen results in the material's heating, local melting, and evaporation, but by no means does complete ablation occur, as only 8% of the entire generated laser energy is absorbed and consequently exploited. As the boiling point requires relatively high amounts of energy to be attained, it is evident that the reflectivity of Al does not allow the target's atoms to acquire an energy that surpasses their cohesion energy, forcing their spontaneous removal from the film.

The reflectance of Al not only affects the extensiveness of the chief irradiated area, but also makes a significant contribution to the form and size of HAZ. The slight discoloration of the pertinent images, right outside the main irradiated region, delimited by the dark gray peripheral circle in each image, indicates that the contact of the laser radiation with the target material is intense enough to locally affect the material's properties.

Another important observation to consider is the higher smoothness and uniformity of the irradiated area in the case of the Al target. As lower amounts of energy are absorbed, in contrast with the Mo thin film, the rays that are strong enough to ablate part of the material are not widely dispersed but more narrowly concentrated around the center of the irradiated area. Moreover, the bulges that appear on the right-hand side of the Al film spots are very small and tend to become nonexistent when the energy falls below a specific value (i.e., 2 mJ). The circular spot that appears to the right of the main irradiated area in Figure 6.9a indicates that large amounts of energy have managed to inflict the removal of numerous aluminum atoms, leading to local ablation. Apparently, as Figure 6.9b and c imply, the incident laser energy of 2 and 1 mJ, respectively, makes this circular spot disappear, while being replaced by a discoloration representing a part of the HAZ. In other words, a 3 mJ laser energy leads to partial ablation of the region in discussion, whereas lower energies are unable to perform the above process. The ostensible smoothness of the final surface does not necessarily suggest that the ablation process on Al is efficient; the factors already described show that performing any manufacturing processes involving laser ablation on Al and even any equally reflective material is not recommended.

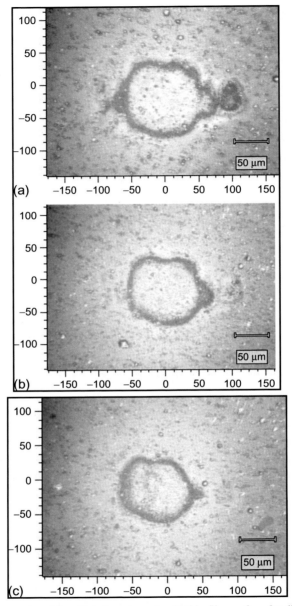

Figure 6.9 Micrographs of the ablated spots on the Al thin film surface for diaphragm diameter of 1.5 μm and incident laser energy values of (a) 3 mJ/pulse, (b) 2 mJ/pulse, and (c) 1 mJ/pulse.

Table 6.2 **Presentation and comparison of important quantities derived from laser ablation experiments on Al and Mo thin films, for diaphragm diameter of 1.5 μm**

Material ID	Laser energy (mJ)	Ablated area (μm²)	Laser fluence (J/cm²)	HAZ area (μm²)	Abl:HAZ area ratio	Ablation efficiency (μm³/mJ)
Mo	3	21,859.8	13.7238	11,106.1	1.96827	2040.2
Mo	2	21,766.3	9.18852	11,392.0	1.91067	3047.3
Mo	1	9937.8	10.0626	13,977.0	0.71101	2782.6
Al	3	15,089.0	19.8820	23,738.0	0.63565	1257.4
Al	2	10,627.6	18.8189	14,517.6	0.73205	743.9
Al	1	7989.4	12.5166	11,424.1	0.69935	878.8

Table 6.2 is of significant interest, as it demonstrates how the different natures of the two materials examined and affect the evolution and final characteristics of the laser ablation process. The cases corresponding to 1.5 μm diaphragm size were isolated, compared, and contrasted in order to discover the different behaviors of the two materials under their irradiation with laser beams.

As Table 6.2 indicates, any attempts to achieve successful results in removing and ablating material prove useless when an Al target is employed. The heating caused to the target because of its irradiation leads to the formation of extensive HAZs, but the material's high index of reflectivity does not favor the removal of particles across large areas, with the eventually ablated regions being over 25% smaller than the respective HAZs. The area ratio between the ablated region and the HAZ is always lower than 1 when Al is the target material, which is consequently unsuitable for any similar type of manufacturing process involving ablation. The target's inappropriateness is also reflected in the significantly low values of ablation efficiency.

By comparing all pairs of spots that correspond to the same laser energy, it can be observed that the substitution of Mo with Al leads to a drop in efficiency that even reaches 75% when the laser energy equals 2 mJ. Moreover, the observation made while studying the ablation of Mo, according to which the efficiency drops when laser fluence increases, also applies when Al is chosen as the target material. However, the two quantities are no longer inversely proportional to each other, something that derives from the high discrepancies in the ablation depth attained during each of the separate cases. This contrasts with the situations referring to Mo being the target, in which the maximum depth of ablation is always found to be the entire thickness of the film. Incidentally, it is very difficult to fully ablate the Al thin film not only because of the target's reflective nature, but also because the film is thicker than the Mo film, generally making the entire ablation process much more difficult with the applied conditions.

The profilometer results indicate that the attainable ablation depth is not always equal to the thickness of the specimen. When the energy of the generated beam is equal to 3 mJ, the rays are capable of locally piercing the specimen up to the other side of its thickness dimension, thus leading to a maximum ablation depth of 390 nm.

When the energies drop to 2 and 1 mJ, the depths managed are, respectively, equal to 350 and 250 nm, something that confirms the beam's inability to perform full surface ablation on the material. In all cases, the highest possible ablation depth values appear only locally on the surface profile, because lower energy rays, generated according to the beam's Gaussian spatial profile, are unable to achieve the same depth of ablation. In addition, the fact that the laser energy decreases as the beam draws deeper inside the target, according to Beer–Lambert's law, should also be taken into account.

In terms of the maximum surface roughness, the respective values corresponding to the 3, 2, and 1 mJ laser energies have been found equal to 1.3135, 1.5463, and 1.0358 µm. These values are much higher than the ones of the Mo target and the same laser energies, something that is justified by the lack of surface uniformity caused mainly because extensive HAZs were formed and expanded during and after the irradiation process. It is also possible that the expulsion of solid and molten material and its movement toward other regions of the thin film, in the form of residues, have contributed to the augmentation of the surface roughness, in contrast with the Mo film.

6.7   Simulation results

The primary MATLAB® computational code was built in such a way that laser irradiation and ablation of the Mo target was simulated, with some key parameters, such as atomic radius, Morse coefficient, macroscopic cohesion energy, etc., being amended for Al. The entire simulation procedure is facilitated by the fact that the materials chosen are generally characterized by a stable crystalline structure, at least for ordinary temperatures up to a 10^4 K order of magnitude.

The cross-section of the Mo specimen is assumed fully rectangular, ignoring any possible geometrical imperfections that require complicated calculations for their simulation. Although it is possible to constrain the simulation process to two dimensions, namely X and Y, because of each film's very low thickness compared to its length and width, such a procedure will not lead to adequate results because the Z dimension is deemed the most important, as being parallel to the direction of all laser rays.

The hypothesis of a theoretically infinite medium is satisfied by employing PBC across the X and Y planes, whereas free and RBC are applied to the upper and lower surface of the Z direction. In order to reduce the computational cost without negatively affecting the precision of calculations, a simple cubic simulation volume containing 15 atoms per dimension (3375 in total) was introduced. Taking into account the interatomic distance of 2.7252 Å in a Mo crystal, it can be concluded that the edge of the formed cube is equal to about 38 Å. Thus, the results obtained from the simulation can be adapted in such a way that they correspond to the actual thickness of each thin film, 280 nm for Mo and 390 nm for Al, assuming a totally homogeneous and isotropic specimen.

In an attempt to simulate the behavior of the material during its irradiation by a single 10 ns laser pulse, this duration was split into 200 isochronous time steps, each being 50 ps long. The number of time steps can be further increased for additional precision, but the choice of 200 steps was still deemed satisfactory, as it allows for the adequate monitoring of the ablation procedure and attribute changes without unnecessarily increasing the calculation time and computational cost. In fact, each separate code constructed for the ablation simulation was able to produce its results throughout the entire 10 ns pulse duration, in less than an hour.

From the first stages of ablation, it is apparent that the purpose of the manufacturing process is the creation of a hole in the specimen, caused by the removal of particles whose contained energy is sufficient to break the cohesion bonds that hold them in the crystalline structure. Atoms not removed after each step move or vibrate at higher velocities without being detached from the simulation volume. An important qualitative observation, while looking at the simulation volume during the first ablation stages, is the decrease of the ablation rate while the respective depth increases. Two reasons justify this statement:

- The Beer–Lambert law is used to model the decay of the laser beam across the Z axis. The number of photons exponentially diminishes according to this law, something that directly affects fewer particles in higher depths.
- It is not necessary for a vaporized particle to exit the crystalline structure. The aerial Mo particles that move inside the simulation volume still absorb photon energy, which, under other circumstances, could have stimulated an increase in the ablation rate.

In Figure 6.10, the evolution of the ablation process is shown, when 10 J/cm² laser fluence is applied. In the same Figure, 22 snapshots of the simulation volume's format during the execution of the code are shown. In Snapshots 2 and 4, the bold lines indicate the regions from where the removal of atoms is initiated. The time moments that correspond to each separate snapshot are shown in Table 6.3, which also gives a representative image of how the number of ablated particles and the temperature of the system evolve over time.

From both Table 6.3 and Figure 6.10, it can be seen that a rise in the system temperature above Mo's melting point coincides with the outset of particle removal from the simulation volume. A nearly linear rise in temperature and number of ablated particles occurs up to 0.9 ns from the simulation start; this moment is the turning point in the ablation rate, which starts to decrease. During the 2.25 ns moment, the maximum system temperature of about 23,000 K is attained; at later stages, the remaining atoms of the simulation volume start to cool. It is also obvious that over 90% of the volume's

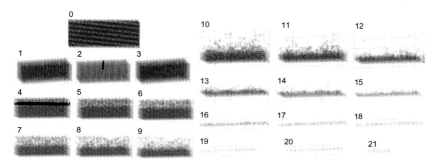

Figure 6.10 Snapshots of laser ablation process evolution for the simulation target, when irradiated by a single 10 ns laser pulse of 10 J/cm² fluence.

Table 6.3 **Presentation of all snapshots shown in Figure 6.10, along with the times to which they correspond, as well as the respective numbers of ablated particles and system temperatures**

Snapshot no.	Simulation time step	Simulation time (ns)	Total number of ablated particles	System temperature (K)
0 (Start)	0	0	0	300
1	7	0.35	24	3271.6
2	8	0.4	76	3676.38
3	9	0.45	145	4106.75
4	10	0.5	230	4551.1
5	12	0.6	517	5400.87
6	14	0.7	774	6305.48
7	16	0.8	1052	7131.34
8	18	0.9	1360	8170.38
9	20	1	1653	8980.71
10	22	1.1	1908	9923.43
11	25	1.25	2261	11,475.87
12	28	1.4	2541	12,532.31
13	32	1.6	2837	15,387.19
14	36	1.8	3032	15,112.57
15	40	2	3160	18,819.54
16	45	2.25	3266	22,846.09
17	50	2.5	3314	11,832.62
18	60	3	3349	7373.97
19	70	3.5	3355	4704.24
20	100	5	3356	6340.41
21 (Finish)	200	10	3369	5473.47

particles have been ablated after 2 ns of laser irradiation, or 20% of the total pulse duration. Thus, for the given fluence of 10 J/cm^2, the high absorptivity of the material combined with the low thickness of the virtual target contribute to the full and rapid ablation across the depth dimension, with the manufacturing process being practically ineffective for the remaining 7–8 ns of the pulse duration.

Figure 6.11 is used to determine the differences in the evolution of the ablation procedure and the mean system temperature between the Mo and Al simulation volumes for the same laser fluence value of 10 J/cm^2. These differences are justified by the different crystalline structures and characteristics of the two materials, as well as the much higher absorptivity of Mo (57.14%) compared to Al (8%). As the absolute energy cohesion value and the melting and boiling points of Al are lower than those of Mo, it is expected that the removal of atoms from the volume will be initiated earlier; the calculations and Figure 6.11 diagrams confirm this statement. Moreover, the temperatures attained on the Al specimen are expectedly low compared to Mo, because the effective fluence of the laser beam, which is absorbed by the target, is approximately seven times lower, taking into account the aforementioned differences in absorptivity.

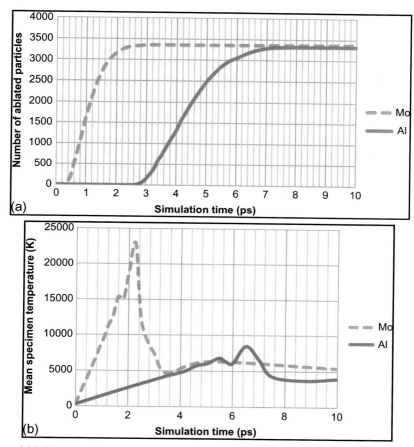

Figure 6.11 MD simulation results on the evolution of (a) the number of ablated particles and (b) the mean specimen temperature during the irradiation of Mo and Al thin films with 10 ns laser pulse of 10 J/cm² fluence.

Finally, it is worth noting that the ablation values calculated as part of both the experimental and MD simulation procedures are in great accordance with each other, with the numerical differences only being minimal. For instance, the efficiency value for a 10 J/cm² fluence is equal to 2782.58 μm³/mJ for the experiment and 2800 μm³/mJ for the simulation, resulting in a slight 0.626% difference. Moreover, the efficiency for a 5.2 J/cm² laser fluence, determined through the experiments, is found to be 5311.74 μm³/mJ, being only 0.72% different from the 5273.78 μm³/mJ value theoretically determined for a fluence of 5 J/cm². The small differences between the experiment and MD simulation are also partially due to the uneven thickness of the film, as being formed through PLD on glass substrates, as well as on the code's independence from the size of the diaphragm through which the laser beam passes before its incidence on the target.

Judging from the conjunctive experimental and theoretical analyses already conducted, three different ablation conditions are observed, depending on the laser fluence applied on the target:

- *Condition A (0.5–0.8 J/cm² laser fluence)*: The region in discussion starts slightly above the 0.4638 J/cm² threshold fluence value, under which thermal influences are almost nonexistent, blocking the material ablation process. For fluences rising above 0.5 J/cm², the ablation process is initially mild, but becomes increasingly stronger, up until the other end of the fluence region (i.e., 0.8 J/cm²). For lower fluences, the main material removal mechanism that takes place is the photothermal one, as the kinetic energy of photons is transformed into thermal, affecting the target's atoms and leading to their fusion and evaporation. Higher fluences favor the appearance of the explosive boiling phenomenon, which involves the formation of an unstable overheated liquid, leading to the spontaneous division of the atom system in the gaseous and stable liquid states. A peak ablation efficiency of about 20,000 μm³/mJ is observed at the higher end of the region.
- *Condition B (0.8–5 J/cm² laser fluence)*: This fluence region ensures the realization of a strong ablation process, during which almost all atoms of the simulation volume are removed. However, as the number of ablated particles does not change significantly as fluence increases, efficiency of the process will start to decrease, reaching somewhat less than 4000 μm³/mJ. Both the photothermal mechanism and explosive boiling phenomenon are encountered throughout this region; the latter appears to an increasing extent as fluence rises. It is possible for some of the atoms to transfer themselves into an ionization state, thus creating an extensive plasma plume over the target, which establishes the plasma shielding effect responsible for the decay of the laser beam. The plasma shielding effect plays a significant role in reducing the ablation efficiency as fluence increases.
- *Condition C (5–20 J/cm² laser fluence)*: In this region, almost all atoms of the simulation volume are ablated, causing the relevant process to be even stronger, principally during the first stages of ablation. However, the fact that fluence has been significantly increased leads to a radical drop in the ablation efficiency, making the process ineffective and fluence values of this region inapplicable for a successful manufacturing process involving laser–material interaction. Explosive boiling and plasma shielding effects are much more intense than in the previous two conditions, directly affecting the efficiency of the process in their turn.

6.8   Conclusions

The process of laser ablation was investigated in detail, both through its direct application on Mo and Al thin films, exploiting their interaction with Gaussian monochromatic laser beams created by an Nd:YAG laser, and through the construction of a thorough 3D MD simulation code in MATLAB®. A comprehensive and reliable methodology has been developed for the study of the ablation process evolution, the number of ablated atoms, and the temperature of the atom system while the process progresses. The results obtained were studied in parallel with the experimental findings, leading to a series of very important results and conclusions, which are the following:

- The MD simulation time can be configured by modifying the number of atoms that constitute the model. The number of particles chosen in each of the model's three dimensions should help the process be represented in the most reliable manner possible, without drastically increasing the computational cost and code execution time whatsoever. In case the desired goal is not achieved, the exportation of results may be significantly delayed; to make matters worse, high volume simulation systems may also produce results severely lacking in precision and realism, something that can be avoided by lowering the MD simulation volume.

- The ablation mechanism and depth, as well as the temperature evolution within the system, depend mainly on the laser fluence of the nanosecond pulse acting on the material. The significant difference between the reflectivity coefficients of Mo and Al may lead to contrasting results as far as the evolution and outcome of the ablation process are concerned.
- The range of laser fluences between 0.5 and $20\,J/cm^2$, investigated during the current analysis, was split into three discrete regions, sharing similar characteristics in process evolution over time, ablation efficiency, and mechanisms that appear.
- Laser fluences examined during the experimental process are relatively high, between 9 and $14\,J/cm^2$ for Mo and 12 and $20\,J/cm^2$ for Al. Taking into account the different absorption coefficients, 57.14% and 8%, respectively, in the two materials, it can be concluded that the Mo thin film of 280 nm thickness used in the experiments is fully ablated across both the depth and a large region of the XY plane. The Al thin film of a 390 nm thickness is only fully ablated when the fluence values are significantly high, close to $20\,J/cm^2$, otherwise the ablation process occurs only partially.
- The improvement of the surface quality during the scribing manufacturing process, using laser pulses, is achieved by decreasing laser fluence. In this way, the ablation process becomes more uniform, providing a smoother and more homogeneous form throughout the whole area of the target's ablated region. Very high fluence values may lead to small sections within the ablated region that have been "burned," locally affecting the color of the target (e.g., its tint becomes brownish in a Mo specimen irradiated with a high fluence laser).
- The ablated specimens, after their experimental irradiation takes place, contain occasional bulges on the left and right sides of the ablated regions. These bulges are favored by higher fluence values, thus deteriorating the surface quality. The bulges appear because the ideally Gaussian laser beam is imperfectly aligned or has been subject to deformation, after its emission from the Nd:YAG laser device and its trajectory through the experimental setup. Due to the above defects in the configuration of the beam, its laser intensity profile across the XY plane detracts from being ideally Gaussian.
- A decrease in the diameter of the diaphragm through which the laser beam passes, causes the incident laser beams to be less dispersed across the XY plane, creating more uniform ablated regions and less extensive HAZs. In this way, the manufacturing process quality improves.
- The nature of the target material affects the correlations between the area of the mainly irradiated/ablated zone and the surrounding HAZs formed after a full pulse has interacted with the target; as Al is a much less absorptive material than Mo, fewer photons are absorbed during the laser irradiation process. Therefore, the ratio between the ablated region and the HAZ is always lower than 1 in Al and mostly higher than 1 in Mo.

References

Adler, B. J., & Wainwright, T. E. (1957). Phase transition of a hard sphere system. *Journal of Chemical Physics*, 27, 1208–1209.

Amoruso, S., Armenante, M., Berardi, V., Bruzzese, R., & Spinelli, N. (1997). Absorption and saturation mechanisms in aluminium laser ablated plasmas. *Applied Physics A*, 65(3), 265–271.

Balandin, V. Y., Niedrig, R., & Bostanjoglo, O. (1995). Simulation of transformations of thin metal films heated by nanosecond laser pulses. *Journal of Applied Physics, 77*(1), 135–142.

Bass, M., Van Stryland, E. W., Williams, D. R., & Wolfe, W. L. (1995). Handbook of optics: Fundamentals, techniques & design *(2nd ed., Vol. 1)*. New York, NY: McGraw-Hill, Inc.

Boyd, I. W. (1992). Photochemical processing of electronic materials. San Diego, CA: Academic Press.

Bozsóki, I., Balogh, B., & Gordon, P. (2011). 355 nm nanosecond pulsed Nd:YAG laser profile measurement, metal thin film ablation and thermal simulation. *Optics & Laser Technology, 43*(7), 1212–1218.

Bulgakov, A. V., & Bulgakova, N. M. (1999). Thermal model of pulsed laser ablation under the conditions of formation and heating of a radiation-absorbing plasma. *Quantum Electronics, 29*(5), 433–437.

Bulgakova, N. M., Burakov, I. M., Meshcheryakov, Y., Stoian, R., Rosenfeld, A., & Hertel, I. V. (2007). Theoretical models and qualitative interpretations of Fs laser material processing. *Journal of Laser Micro/Nanoengineering, 2*(1), 76–86.

Cheng, C., & Xu, X. (2005). Mechanisms of decomposition of metal during femtosecond laser ablation. *Physical Review B, 72*, 165415.

Chiang, K.-N., Chou, C.-Y., Wu, C.-J., & Yuan, C.-A. (2006). Prediction of the bulk elastic constant of metals using atomic-level single-lattice analytical method. *Applied Physics Letters, 88*, 171904.

Conde, J. C., Lusquiños, F., González, P., Serra, J., León, B., Dima, A., et al. (2004). Finite element analysis of the initial stages of the laser ablation process. *Thin Solid Films, 453–454*, 323–327.

Fang, R., Zhang, D., Li, Z., Yang, F., Li, L., Tan, X., et al. (2008). Improved thermal model and its application in UV high-power pulsed laser ablation of metal target. *Solid State Communications, 145*(11–12), 556–560.

Gamaly, E. G., Madsen, N. R., Duering, M., Rode, A. V., Kolev, V. Z., & Luther-Davies, B. (2005). Ablation of metals with picosecond laser pulses: Evidence of long-lived nonequilibrium conditions at the surface. *Physical Review B, 71*, 174405.

Gordon, P., Balogh, B., & Sinkovics, B. (2007). Thermal simulation of UV laser ablation of polyimide. *Microelectronics Reliability, 47*(2–3), 347–353.

Leitz, K.-H., Redlingshöfer, B., Reg, Y., Otto, A., & Schmidt, M. (2011). Metal ablation with short and ultrashort laser pulses. *Physics Procedia, Part B, 12*, 230–238.

Li, L., Zhang, D., Li, Z., Guan, L., Tan, X., Fang, R., et al. (2006). The investigation of optical characteristics of metal target in high power laser ablation. *Physica B: Condensed Matter, 383*(2), 194–201.

Liu, D., & Zhang, D.-M. (2008). Vaporization and plasma shielding during high power nanosecond laser ablation of silicon and nickel. *Chinese Physics Letters, 25*(4), 1368–1371.

Miller, J. C. (1994). Laser ablation. Principles and applications. Berlin, Heidelberg: Springer Verlag.

Nedialkov, N. N., & Atanasov, P. A. (2006). Molecular dynamics simulation study of deep hole drilling in iron by ultrashort laser pulses. *Applied Surface Science, 252*(13), 4411–4415.

Nedialkov, N. N., Imamova, S. E., Atanasov, P. A., Berger, P., & Dausinger, F. (2005). Mechanism of ultrashort laser ablation of metals: Molecular dynamics simulation. *Applied Surface Science, 247*(1–4), 243–248.

Ohmura, E., Fukumoto, I., & Miyamoto, I. (2000). Molecular dynamics simulation of ablation process with ultrashort-pulse laser. In *Proceedings of SPIE: Vol. 4088. First international symposium on laser precision microfabrication, 84.*

Oliveira, V., & Vilar, R. (2007). Finite element simulation of pulsed laser ablation of titanium carbide. *Applied Surface Science*, *253*(19), 7810–7814.

Pronko, P. P., Dutta, S. K., Du, D., & Singh, R. K. (1995). Thermophysical effects in laser processing of materials with picosecond and femtosecond pulses. *Journal of Applied Physics*, *78*(10), 6233–6240.

Schäfer, C., Urbassek, H. M., & Zhigilei, L. V. (2002). Metal ablation by picosecond laser pulses: A hybrid simulation. *Physical Review B*, *66*, 115404.

Schommers, W. (1986). Molecular dynamics and the study of anharmonic surface effects. In W. Schommers & P. von Blanckenhagen (Eds.), *Topics in current physics: Vol. 41. Structures and dynamics of surfaces I* (pp. 199–243). Berlin, Heidelberg: Springer.

Shirk, M. D., & Molian, P. A. (2001). Ultra-short pulsed laser ablation of highly oriented pyrolytic graphite. *Carbon*, *39*(8), 1183–1193.

Simon, P., & Ihlemann, J. (1996). Machining of submicron structures on metals and semiconductors by ultrashort UV-laser pulses. *Applied Physics A*, *63*(5), 505–508.

Stavropoulos, P., Stournaras, A., Salonitis, K., & Chryssolouris, G. (2010). Experimental and theoretical investigation of the ablation mechanisms during femtosecond laser machining. *International Journal of Nanomanufacturing*, *6*(1–4), 55–65.

Stillinger, F. H., & Weber, T. A. (1985). Computer simulation of local order in condensed phases of silicon. *Physical Review B*, *31*(8), 5262–5271.

Vasantgadkar, N. A., Bhandarkar, U. V., & Joshi, S. S. (2010). A finite element model to predict the ablation depth in pulsed laser ablation. *Thin Solid Films*, *519*(4), 1421–1430.

Yamashita, Y., Yokomine, T., Ebara, S., & Shimizu, A. (2006). Heat transport analysis for femtosecond laser ablation with molecular dynamics-two temperature model method. *Fusion Engineering and Design*, *81*(8–14), 1695–1700.

Zhigilei, L. V., Kodali, P. B. S., & Garrison, B. J. (1997). Molecular dynamics model for laser ablation and desorption of organic solids. *Journal of Physical Chemistry B*, *101*(11), 2028–2037.

Zhigilei, L. V., Lin, Z., Ivanov, D. S., Leveugle, E., Duff, W. H., Thomas, D., et al. (2010). Atomic/molecular-level simulations of laser–materials interactions. In A. Miotello & P. M. Ossi (Eds.), *Laser–surface interactions for new materials production* (pp. 43–79). Berlin, Heidelberg: Springer.

Manufacturing processes of shape memory alloys

7

A.P. Markopoulos, I.S. Pressas, D.E. Manolakos,
National Technical University of Athens (NTUA), Athens, Greece

7.1 Introduction to SMAs

Shape memory materials have recently received a wide acknowledgment in the scientific community, mostly because of their increased use in biomedical and robotic applications. A lot of research has been conducted in order to define the mechanisms of superelasticity (SE) and shape memory effect (SME) that give these materials their characterization as smart materials. Most of the research is referred to nickel and titanium intermetallic compounds, also known as NiTi or Nitinol alloys, which occupy the majority of the shape memory products' market.

7.1.1 Superelasticity

Superelasticity (also termed pseudoelasticity) refers to a characteristic that certain materials possess. The superelastic materials can return to an initial shape when the applied deformational stress is removed. When the stress is removed in such a material, it leads to a subsequent recovery of the deformation strain, thus the material returns to its original shape. This phenomenon is attributed to the reversible transformation occurring from the austenite phase to the "parent" mechanically formed martensite, also termed stress-induced martensite. The transformation caused by the applied mechanical load usually occurs in a nominal constant temperature above A_f.

In order to better clarify the SE mechanism, Figure 7.1, depicting a stress–temperature curve and a stress–strain curve, is presented. The temperatures shown in the stress–temperature curve are characteristic for each material and are described below:

- M_s: martensite phase transformation start temperature upon cooling.
- M_f: martensite phase transformation finish temperature upon cooling.
- A_s: austenite phase transformation start temperature upon heating.
- A_f: austenite phase transformation finish temperature upon heating.

As shown from the curves in Figure 7.1, the material from point (A) is stressed at a constant temperature, while being in a stable austenite phase. The resulting deformation is elastic until a certain point (B), where the material reaches the state that the martensitic transformation begins. From this point on, the transformation that takes place is accomplished under a constant stress, while the strain continues to increase, until a maximum strain level (point C). The maximum strain level varies according to the material. The curved section between points (B) and (C) is termed the stress "plateau." At this point, the phase transformation from austenite to martensite is completed and

Materials Forming and Machining. http://dx.doi.org/10.1016/B978-0-85709-483-4.00007-7

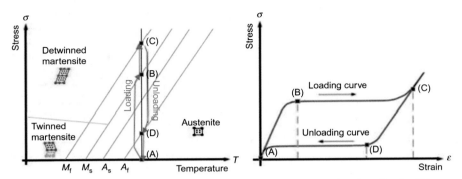

Figure 7.1 Stress–temperature (left) and stress–strain (right) curves that describe the SE behavior.

the curve that describes the behavior of the material is different. This new behavior usually presents a small temperature hysteresis ($\Delta T = A_f - M_s$), whereas the "parent" martensite interfaces present some mobility. In should be noted that the level of the stress "plateau" is heavily dependent on the applied temperature (Bellouard, 2002). Further stressing the material from point (C) will only lead to elastic deformations of the detwinned martensite. Finally, after the stress is removed, the material begins to return to its stable austenite phase, until it fully transforms this phase (point D), thus the cycle can be repeated.

SE is different from the regular elasticity of bulk materials, as the mechanisms of the two phenomena differ. Unlike the mechanism of the SE described above, the mechanism of the regular elasticity is attributed to the variation of the interatomic spacing within the material, as described by Hooke's law. Typical values of regular elasticity can be up to 0.5% for most bulk materials, whereas the value of SE in Fe-based shape memory alloys (SMAs) can be up to 13%. In single crystal materials, it can even reach values of 15% or more (Bellouard, 2002).

7.1.2 Shape memory effect

The SME refers to the recovery of plastic strains in an initially deformed low-temperature material, through the mechanism of austenite–martensite phase transformation. The shape which the material returns to, after the phase transformation, can be "memorized" by the material through its deformation in high temperature (austenite phase). Thus, a material deformed in the martensite phase (deformed below the A_s temperature) can recover its plastic strain and return to its high-temperature shape via heating to the austenite phase. In other cases, the material can return to its low-temperature shape from the austenite phase via cooling to the martensite phase.

In order to clarify the SME mechanism, Figure 7.2 is given. In this graph, a thermo-mechanical loading path in a stress–strain–temperature space is presented. It is worth noticing that the below described phenomenon is the simplest case of the SME, termed as one-way shape memory effect or SME for short. The data for the graph of Figure 7.2 represent the SME mechanism of a typical NiTi wire, tested under uniaxial loading.

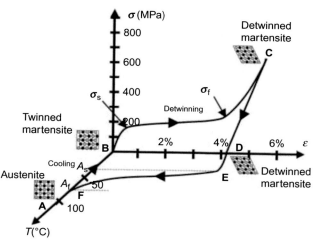

Figure 7.2 Uniaxial loading path of a typical NiTi SMA sample, shown in a stress–strain–temperature space (Kumar & Lagoudas, 2008).

As shown in the graph, the test begins with the material being in the austenite or "parent" phase (point A). From this point, the material is cooled to a temperature below the material's forward transformation temperatures (M_s and M_f), where a twinned martensite is formed (point B). After the material is completely in the martensite phase, the uniaxial loading begins. The applied load continues to advance until the start stress level (σ_s) is surpassed. Beyond this point, the grains of the material begin to reorientate, leading to certain martensitic variants with a favorable orientation to grow at the expense of other less favorable variants. It should be noted that the stress needed for this process is far lower than the plastic yield strength of the martensite. This process, termed detwinning, is completed at the final stress level (σ_f), corresponding to the end of the stress "plateau" (point C). The next step is the elastic unloading of the material, where its state remains unchanged (point D). After the material is completely unloaded, a heating process leads the material to begin transforming to austenite. This process begins above the material's A_s temperature (point E) and it is completed when the material reaches a temperature above its A_f temperature (point F). With only the "parent" austenitic phase remaining, the material regains its original shape (point A), due to the absence of the permanent plastic strain of the detwinned martensitic phase. The strain recovered as a result of the martensite to austenite phase transformation is termed the transformation strain (ε^t). Cooling the material from this point will result in a twinned martensitic phase, which will bring no change to the shape of the material, thus the above explained cycle can be repeated (Kumar & Lagoudas, 2008).

In most cases, the trigger that causes the phase transformation is heating the material above its A_f temperature (thermo-responsive SME). However, this is not the only case; other possible effects that can cause this transformation are the direct application of light (photo-responsive SME), the exposure of the material to a certain chemical substance (chemo-responsive SME), the application of a mechanical load (pressure-responsive SME), or the application of a magnetic field (magnetic-responsive SME) (Huang et al., 2013).

7.1.3   Transformation-induced fatigue in SMAs

The repeating thermo-mechanical cycle described above, either to induce SE or a thermal phase transformation in a SMA, both under applied load, can lead to a premature fatigue of the material. The fatigue behavior of a SMA is generally attributed to a variety of factors, like the processing of the material (casting, manufacturing processes, heat treatment, etc.), the working conditions (stress, strain, thermal transitions, etc.), environmental factors (temperature, humidity, etc.), and transformation-induced microstructural modifications (e.g., defects on grain boundaries due to strain incompatibilities). By repeatedly loading the material, both mechanically and thermally, microstructural changes occur. These microstructural changes gradually degrade the SMA behavior.

As explained, the SMA fatigue can be attributed both to the mechanical and the thermal loading. Mechanically loading the material can either induce a complete phase transformation (starting from a complete martensite phase and finishing with a complete austenite phase) or a partial transformation (the starting or the finishing point can be consisting of a mixture between the two phases). Depending on the magnitude of the applied stress, this can lead to a varying fatigue life (e.g., about 10^3–10^4 cycle repeats for high load levels, about 10^7 cycle repeats for low load levels) within the elastic regime. Similarly, thermal loading can highly affect the overall fatigue life of a SMA. This is mainly due to the amount of transformation strain that occurs during each cycle (i.e., complete or partial transformation), as well as the maximum stress level applied during each cycle. During a partial transformation, the martensite transformation is not complete, thus the associated transformation strain is limited. This leads to a higher fatigue life of a SMA.

Other factors that can heavily affect the fatigue life of SMAs are microstructural factors such as the crystallographic orientation or the precipitate size. Most of these factors can be neutralizing through heat treating processes in specific conditions. In needs to be noted though, that nonoptimal conditions can induce certain processes such as oxidation or corrosion, which will subsequently lead to a faster crystallographic deterioration and thus a lower fatigue life for the SMA (Kumar & Lagoudas, 2008).

7.2   Deformation mechanisms

In forming manufacturing processes, high stress is applied to the material in order to induce permanent plastic strain, thus giving it a desirable shape. The overall process may include a combination of many different loading modes to the material. For this reason the basic loading modes are described below. In order to ascribe isothermal mechanical behavior one should consider a behavior similar to the elastoplastic one. Such an approximation is presented in Figure 7.3. It should be noted, that this

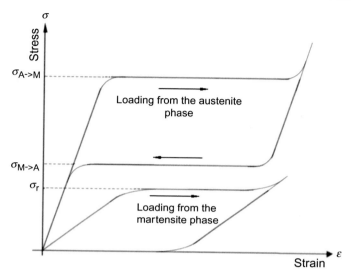

Figure 7.3 Isothermal elastoplastic model of the SMA mechanical characteristics.

approximation is the same for both superelastic and martensitic curves. For the former, the critical stress refers to the stress "plateau," while for the latter, the critical stress is the one required for inducing reorientation of the martensitic variants above A_f temperature.

7.2.1 Tension

The most basic loading mode is tension. In this mode, the material is axially loaded and the stress is distributed uniformly inside the volume of the material. Thus, the material is uniformly transformed, resulting in the optimal efficiency in transformation energy. However, the range in which the material can move is fairly low. It should be noted that in this loading mode the normal stress is directly proportional to the applied force. The behavior of a SMA in tension mode is shown in the material's characteristic stress–strain curve. In other words, the curve that characterizes the superelastic behavior during the loading cycle also presents the material's behavior in tension (also see Section 7.1.1).

7.2.2 Bending

In this mode the distribution of stress across the material is not constant. Figure 7.4 represents a SMA strip subjected to pure bent. According to the classic bending analysis, there are three different regions in a bent material. These consist of a tended (outer) region, a compressed (inner) region, and an elastically deformed (middle) region. The first two regions are subjected to an applied stress that surpasses the critical value, above which the material begins to transform into stress-induced martensite. This results in a combination of elastic and ideally plastic deformation taking place in these

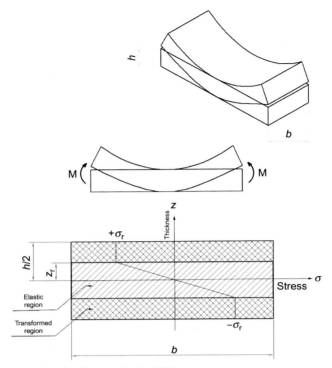

Figure 7.4 Pure bending loading mode of a NiTi strip.

regions. On the other hand, the middle region is subjected to a stress level below the above-mentioned critical value, thus the deformation in this region presents a pure linear-elastic behavior. It should be noted that the behavior expected upon loading can be approximated by elastoplastic-like behavior, with the tended region presenting the same behavior as the compressed region.

Nonetheless, research by Liu, Xie, Van Humbeeck, and Delaey (1998) showed that this approximation is not representative of the behavior of a NiTi alloy, as the compression region does not present a stress "plateau." Even still, this approximation can present some rough results, which can be improved with the addition of a strain hardening condition. In Figure 7.4, the below explained correspondence is in effect:

- h: the strip's thickness
- b: the strip's width
- σ_r: the critical stress-induced martensite transformation stress
- R: the radius of the curvature due to bending
- z: the relative thickness measured from the center line of the strip
- z/R: the strain of the strip due to bending
- z_r: the relative thickness with a stress level equal to σ_r
- E: the material's Young modulus

If the material is in the austenite phase, the applied stress can be expressed by the following equation:

$$\sigma(z) = \begin{cases} E\dfrac{z}{R}, & z \le z_r \\ \sigma_r, & z > z_r \end{cases} \tag{7.1}$$

where $-\dfrac{h}{2} \le z \le \dfrac{h}{2}$.

Subsequently, the applied moment M can be expressed by the following equation:

$$M = \int_A z\sigma(z)\,dA = \frac{2}{3}\frac{Eb\sigma_r^3}{R} + b\sigma_r\left(\frac{h^2}{4} - z_r^2\right) \tag{7.2}$$

In the last equation, the term $\dfrac{2}{3}\dfrac{Eb\sigma_r^3}{R}$ expresses the behavior of the elastic region, while the term $b\sigma_r\left(\dfrac{h^2}{4} - z_r^2\right)$ expresses behavior of the transformed region of the material.

Finally, if the relative thickness of the boundary layer z_r is assumed, where

$$z_r = \sigma_r \frac{R}{E}$$

proved by equating the two terms of Equation (7.1), then the expression of the bending moment is written as a function of the curvature radius R, according to the following equation:

$$M(R) = \begin{cases} \dfrac{bh^2}{4}\sigma_r - \dfrac{b\sigma_r^3 R^2}{3E^2}, & R \le \dfrac{Eh}{2\sigma_r} \\ \dfrac{bh^3}{12}\dfrac{E}{R}, & R > \dfrac{Eh}{2\sigma_r} \end{cases} \tag{7.3}$$

7.2.3 Torsion

Similar to bending, the stress distribution across the material in torsion is not constant. As shown in Figure 7.5 where a cylindrical sample is subjected to torsion, the center of the material deforms elastically, while the maximum deformation occurs in the material's peripheral surface. If the following correspondence is in effect:

- M_t: the torsional moment
- τ: the shear stress
- G: the material's shear modulus
- R: the radius of the sample
- r: the radius along the cross section
- θ: the twist angle per unit of length

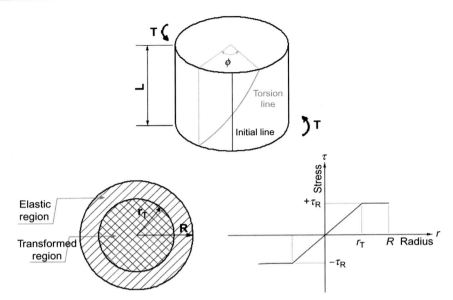

Figure 7.5 Pure torsion loading mode.

- τ_R: the critical stress-induced martensite transformation shear stress
- φ: the total twist angle
- L: the total length of the sample

then the shear stress in elastoplastic behavior is expressed by the following equation:

$$\tau = \begin{cases} Gr\theta, & 0 \le r \le r_T \\ \tau_R, & r_T \le r \le R \end{cases} \tag{7.4}$$

Thus, the torsion moment is expressed by equation:

$$\Gamma = \int_A \tau r dA = 2\pi \int_0^R \tau r^2 dr = 2\pi \left[\int_0^{r_T} (Gr\theta) r^2 dr + \int_{r_T}^R \tau_r r^2 dr \right] \tag{7.5}$$

Similar to the above, in this last equation, the first term expresses the behavior of the material's elastic region, while the second term expresses the behavior of the material's transformed region. Finally, Equation (7.5) can be expressed as a function of the total twist angle, as:

$$\Gamma(\varphi) = \begin{cases} \dfrac{\pi r^4}{2} \dfrac{G\varphi}{L}, & \varphi \le \dfrac{\tau_R L}{GR} \\[4mm] 2\pi \left(\tau_R \dfrac{R^3}{3} - \dfrac{1}{12} \dfrac{L^3 \tau_R^4}{G^3 \varphi^3} \right), & \varphi > \dfrac{\tau_R L}{GR} \end{cases} \tag{7.6}$$

7.3 Manufacturing processes

In this section, the basic fabrication methods of SMAs are discussed. The sequence in which the methods are presented matches an actual complete fabrication process, with the manufacturing processes described last.

7.3.1 Casting

The majority of the SMA products are cast in a semifinished form, according to the intended application. The most common forms include ingots, slabs, billets, ribbons, foils, rods, or wires. The desired form of the cast product defines the casting method used. The above-mentioned forms are mainly produced through continuous casting methods with rapid solidification, such as melt-spinning, planar flow casting, and twin-roll casting (Lojen, Gojić, & Anžel, 2013). The tools used for each fabrication process are a combination of the desired form of the semifinished product, the SMA used and the mechanism of the chosen casting process. One should carefully research the ideal melting and casting method for the SMA used, as well as the process that best delivers the desired form (Funakubo, 1987; Ito, Sahara, Farjami, Maruyama, & Kubo, 2006; Lojen et al., 2005).

The melt is prepared by melting each of the individual components separately, in the appropriate stoichiometric proportions. A small differentiation in the quantity of each of the melt's components results in a different SMA with great differentiation in its properties. The same applies for the addition of certain third elements, even if third elements are usually added in small concentrations. In order to better clarify the effect of a minimal change in a SMA's component concentrations, Table 7.1 is presented with some characteristic Nitinol SMAs, as well as their characteristic temperature levels. When different elements are in the correct concentrations, they are mixed in a single mixture and then cast, via an appropriate casting method, in the desired form. In some cases, the initial casting of the SMA may be followed by several remelts, in order for the resulting material to present better homogeneity (Wu, 2001).

Another great threat to the integrity of the alloy is the existence of some contaminating factors. Contaminants can enter the alloy either because of the low purity of the original materials, or during the casting of the product. Using component materials of high purity (usually over 99.9%) can eliminate the first case (Ito et al., 2006; Lojen et al., 2013). In order to prevent possible contamination of the alloy during its fabrication, both the melting and the casting of the component materials are usually performed under a protective environment, consisting of Ar or other noble gas, or in a vacuum. Other sources of contaminating elements during this step are the crucible used and the oxides formed during the melt's solidification. This makes the need for proper selection of a noncontaminating crucible, suitable for the melt, extremely crucial. In most cases, the crucible can be made out of graphite, calcia, alumina, magnesia, and other similar material, with some being more suitable than others, depending on the casted alloy. Apart from the protective atmosphere, in order to prevent the formed oxides from contaminating the main volume of the final material, sometimes water is circulated around the cast material. The circulated water increases the heat

Table 7.1 **Transformation temperatures of different NiTi-based SMAs**

NiTi-based SMAs	M_f	M_s	A_s	A_f	References
$Ti_{50}Ni_{50}$	15	55	80	89	Lidquist and Wayman (1990)
$Ti_{49.5}Ni_{50.5}$	−78	−19	9	53	Strnadel, Ohashi, Ohtsuka, Ishihara, and Miyazaki (1995)
$Ti_{49}Ni_{51}$	−153	−114	−89	−40	Funakubo (1987)
$Ti_{49}Ni_{41}Cu_{10}$	8	30	35	50	Strnadel et al. (1995)
$Ti_{50}Ni_{40}Cu_{10}$	21	41	53	67	Strnadel et al. (1995)
$Ti_{44}Ni_{47}Nb_9$	−175	−90	−85	−35	Zhao et al. (1990)
$Ti_{42.2}Ni_{49.8}Hf_8$	50	69	111	142	Potapov et al. (1997)
$Ti_{40.7}Ni_{49.8}Hf_{9.5}$	61	90	118	159	Potapov et al. (1997)
$Ti_{40.2}Ni_{49.8}Hf_{10}$	103	128	182	198	Potapov et al. (1997)
$Ti_{35.2}Ni_{49.8}Hf_{15}$	95	136	140	210	Potapov et al. (1997)
$Ti_{30.2}Ni_{49.8}Hf_{20}$	127	174	200	276	Potapov et al. (1997)
$Ti_{48}Ni_{47}Zr_5$	20	65	75	138	Hsieh and Wu (1998)
$Ti_{43}Ni_{47}Zr_{10}$	45	100	113	165	Hsieh and Wu (1998)
$Ti_{38}Ni_{47}Zr_{15}$	100	175	175	230	Hsieh and Wu (1998)
$Ti_{33}Ni_{47}Zr_{20}$	205	275	265	330	Hsieh and Wu (1998)
$Ti_{50}Pd_{50}$	550	563	580	591	Lidquist and Wayman (1990)
$Ti_{50}Ni_{20}Pd_{30}$	208	241	230	241	Lidquist and Wayman (1990)
$Ti_{50}Ni_{10}Pd_{40}$	387	403	419	427	Lidquist and Wayman (1990)
$Ti_{50}Ni_5Pd_{45}$	467	486	503	509	Lidquist and Wayman (1990)
$Ti_{50}Ni_{45}Pt_5$	10	29	36	49	Lidquist and Wayman (1990)
$Ti_{50}Ni_{40}Pt_{10}$	−8	18	−27	36	Lidquist and Wayman (1990)
$Ti_{50}Ni_{30}Pt_{20}$	241	300	263	300	Lidquist and Wayman (1990)
$Ti_{50}Ni_{20}Pt_{30}$	537	619	626	702	Lidquist and Wayman (1990)

flow from the outer surface of the material, causing the oxides to be formed there, thus maintaining the purity of the alloy while making the oxides easier to remove afterwards. Contaminations can have significant effects in the quality of the final material. In most cases, they cause significant changes in the temperature levels of the final material, making its behavior unpredictable, or even incompatible for the intended application (Olier et al., 1997). Moreover, due to the highly reactive behavior of some components (e.g., Ti), melting processes in a vacuum are often necessary. Some such common processes are vacuum induction melting and the vacuum consumable arc melting processes (Wu, 2001).

Finally, one of the most important steps of the casting process in the fabrication of SMAs is the solidification of the melt. As mentioned above, most of the processes used in the casting of a SMA, are rapid-solidification processes, which means that the heat flow from the melt is significantly high (Lavernia & Srivatsan, 2010). This causes the creation of a specific microstructure in the final solid, something important in order for the final material to present the SME (Lojen et al., 2013). The high heat flow from the melt is usually achieved, either by circulating a cooling medium around the exit of the mould, or by quickly drenching the mould with the melt in the cooling medium. Common cooling mediums are water, brine solutions, and liquid nitrogen. The problem of the proper solidification method and the appropriate cooling medium for each casting process needs to be carefully addressed. A very high heat flow rate may result, depending on the casted SMA, in a thorough martensite phase, but it may also lead to the formation of defects (e.g., cracks). Finally, certain rapid-solidification techniques lead to a final material with low homogeneity. This lack of homogeneity is usually improved via heat treatment processes.

7.3.2 Heat treatment

Heat treating refers to a variety of processes through which the material microstructure is being altered. Via the microstructure alteration, many physical and chemical properties of the material change. For the SMAs, heat treatment processes present an important role and are necessary for the SME mechanism to take place.

One of the most well-known heat treating processes used during the fabrication of a SMA involves the annealing of the material right after the casting process. Depending on the casting process, the microstructure of the SMA right after the casting process may present a lack of homogeneity. This affects several of the mechanical and physical properties of the SMA, such as its ductility, its hardness, and its strain recovery percentage. During the annealing process, the as-cast material is placed inside a heat treating oven, where it remains for several hours, while being heated above a certain temperature and at a certain pressure. The resulting material presents a more homogenous microstructure, which, depending on the annealing time, might improve in a single phase of the crystal growth. This effect poses a significant role in the ductility of the SMA, which defines how easily the material can be further manufactured, while it may also change the A_s temperature level of the material (Miyazaki, Kohiyama, Otsuka, & Duerig, 1990; Xu et al., 2003). The same process is also applied after the material has been subjected to cold manufacturing processes in order to decrease the residual stresses induced. This results in relieving the work-hardening of the material, rendering the material ready for further processing (Miller & Lagoudas, 2001). If the material is submitted to several manufacturing processes or to a large deformation through multiple passes, the heat treatment process can be performed in several steps, usually between each deformation pass (Wu, 2001).

Perhaps the most important role of the heat treating processes in the fabrication of a SMA, is the setting of the austenite shape in the material, a procedure often termed "training." Usually, training of a SMA involves the annealing of the material above a certain temperature level (A_f), while in the desired shape, with a subsequent quench in a cooling medium. The effect of this process is the rapid transformation from austenite to martensite, with the austenite form altered to match the new shape. In order for the material to maintain the desired shape throughout this process, the product needs to be firmly constrained during the annealing, or it may partly recover its initial shape. The time and temperature level of this process vary, depending on the material in use. Generally, the temperature has to be above the A_f temperature of the material, while the duration has to be long enough that all of the material's volume reaches thermal equilibrium with its environment. Usually, all of the necessary manufacturing processes precede the material's shape setting. This is due to the fact that the shape setting of a SMA needs to be performed in the final product so that the strain recovery can be fully controllable. What is more, if a deformation manufacturing process is cold, the annealing performed for the shape setting will relieve some of the residual stresses induced from the manufacturing process, thus decreasing the material's work-hardening (Miller & Lagoudas, 2001).

Annealing can also be used to recover a percentage of the transformation-induced fatigue of the SMAs. As mentioned previously (see Section 7.1.3), the repeated thermo-mechanical loading cycles to which a SMA is submitted during its lifetime, cause permanent microstructural changes in the material. These changes lead gradually to a degradation of the shape memory behavior. As a result, annealing of the material can revert some of these permanent microstructural changes, thus restoring the shape memory behavior of the material (Kumar & Lagoudas, 2008). Attention is needed, though, so that the annealing process will not change the initial properties of the SMA (e.g., the material's temperature levels).

In some SMAs, apart from the annealing process described above, an aging process may also be needed. Through the process of aging, the finish austenite temperature level A_f of a SMA can be significantly increased. In most cases this is due to the precipitation of certain areas within the volume of the material that, because of their different orientation, led to a lower A_f temperature level. The process involves the firm constraint of the material and the subsequent heating at a certain temperature. After that, a quick drench in a cooling medium sharply defines the above-mentioned temperature level. The main difference between annealing and aging lies in the duration and the maximum temperature of each process. In general, for the same material, aging is performed in lower temperatures than the annealing, but for an extended period of time.

Common methods of heat transition in heat treatment processes involve a protective atmosphere or a vacuum furnace, salt or sand baths, usage of heated dies, and other heating methods. The actual temperatures for a given material, in which the heat treating should be performed, must be carefully selected. Finally, common methods of restraining the SMA product before heat treating involve the use of certain devices, such as a fixture, a mandrel, or others. Choosing the right constraining device is heavily affected by the initial form given during casting, as well as the heating device used.

7.3.3 Forming

After the heat treatment of the material, several manufacturing processes are often used to achieve the desired product dimensions. These manufacturing processes can either be cold or hot. Although hot-worked products present better behavior in terms of their thermo-mechanical properties, sometimes cold manufacturing processes are needed. As noted above, in case a SMA is manufactured via cold processes, the material should be submitted afterwards to annealing to reduce the work-hardening. Furthermore, both in cold and in hot processes, if the deformation is high, relative to the product's initial dimensions, the overall deformation is performed in various passes in order to prevent the possibility of defect formation.

A SMA's workability in deformation manufacturing processes is highly connected to the material's ductility. The more ductile a material, the easier it can be successfully manufactured through deformation. Comparison between the different types of SMA shows that Cu-based alloys are generally more ductile than Ti-based alloys, a fact that makes forming manufacturing processes more appropriate in Cu-based alloys (ASM, 1993). Still, Ti-based SMA product fabrication through deformation processes is also common. Generally, a SMA's ductility is heavily affected by certain factors, such as the concentration of a specific element in the alloy, the temperature of the process, and whether the alloy is subjected to intermediate heat treatment or not (Miyazaki et al., 1990).

A major problem that forming manufacturing processes pose, in terms of the material's strain recovery, rests with the main mechanism of these processes. During a deformation process, the material is forced to pass through a certain opening, or to occupy a certain volume. This procedure gives different orientation to the material grains, thus resulting in a microstructure with different grain orientations (Leclercq & Lexcellent, 1996). As discussed above, this phenomenon causes a decrease in the material's strain recovery, thus resulting in a fatigued SMA (see also Section 7.1.2). This phenomenon is much more dominant in cold manufacturing processes, as during a hot process the material is dynamically recrystallized simultaneously (Cong et al., 2006). Annealing the material after a deformation manufacturing process can change the orientations of the grains to more favorable ones (Kumar & Lagoudas, 2008).

In terms of tools and machinery, each manufacturing process presents different requirements. The tools are usually constructed of hard materials, as a lot of the SMAs require high loads in order to be deformed. For example, in the case of Ti–Ni alloys, carbide and diamond tools are commonly used due to the alloys' increased hardness. Similarly, several consumable components needed in forming manufacturing processes (e.g., lubricants) vary according to the SMA manufactured. For instance, in the case of NiTi alloys, common lubricants are oil-based lubricants, water-based lubricants with the addition of graphite, sodium stearate soap, molybdenum disulphide, and others (Wu, 2001; Wu, Lin, & Yen, 1996). The optimal tools and consumables that should be used, according to the desired manufacturing process and SMA, need to be appropriately selected.

In general, forming processes in SMAs present similar behavior to those of alloys of the same basic component, for example, the manufacturing of Nitinol is fairly similar to that of an ordinary Ti-based alloy. For this reason, the workability of similar alloys to those of interest in a specific manufacturing process may be investigated before proceeding to the actual process. The most common deformation manufacturing processes are presented in the following paragraphs.

7.3.3.1 Forging

Forging is a forming process in which the metal is forced to occupy a given geometry. Usually, the applied load is uniaxial, given by a press or a hammer. The geometry which the material is forced to occupy is engraved in one or both of the dies, according to the forging type. In most cases, this geometry is near the desired shape of the material. Due to the nature of the manufacturing process, most of the material deforms plastically, while a small percentage of the material deforms elastically. So, in order to achieve the exact dimensions of the final product, the dies are designed undersized or alternatively, the product may proceed to a machining manufacturing process, in order to be formed in the desired dimensions (ASM, 1993).

Forging is a common process for the manufacturing of SMA ingots or billets. As mentioned above, due to the fact that the SMA has to flow in certain directions in order to occupy the given geometry of the dies, the resulting material will present a microstructure with varying orientations. In order to decrease this phenomenon, forging is usually performed as a hot process, as during a hot forging the material presents finer grains, caused by the dynamic recrystallization occurring simultaneously during this process (Cong et al., 2006). It should be noted that performing hot forging in a SMA requires careful design of the dies, both in the aspect of geometry and material. In that way, the dies used will present high durability in the mechanical and thermal loading, thus a long lifetime will be ensured.

Another way to reduce the negative effects of hot forging is by using an outer covering material during the process. This covering material, also termed a "jacket," receives the majority of the work-hardening occurring during the process, something that would normally be induced into the SMA. After the material is forged to its final dimensions, the jacket is removed so the protected SMA underneath has significantly improved properties with little need for further heat treating. Furthermore, the surface of the final product is better, without the need for subsequent machining.

In case of cold forging, the material's strain rate is higher, causing more unstable deformations. This results in a lower quality SMA, while the possibility of defect formation and decreased surface finish is a lot higher (Yeom et al., in press). In order to acquire a higher quality SMA, frequent inter-pass annealing is considered vital (Wu, 2001). Furthermore, frequent inter-pass annealing as well as performing the manufacturing process in many low strain deformation passes, can significantly lower work-hardening and as a result the possibility of defect formation. Another common method employed in cold forging involves the cover of the material by a thin oxide layer. This layer usually consists of zirconia, yttria, or another oxide, which acts as a lubricant while it also absorbs the direct applied loads, thus protecting the underneath material.

Forging in SMAs is usually employed in order to form complicated forms or as an intermediate manufacturing process. Complicated forms are attained through impression-die (closed-die) forging. In this type of forging, lubrication may be used, although it is not always necessary. On the other hand, if the result of the forging process is an intermediate product, an open-die forging is commonly employed. Common intermediate forms of SMAs are rods and billets, as well as several rotational forms.

One should properly research the ideal parameters of a forging process for a chosen SMA. Cong et al. (2006) showed that according to the SMA used and the type of the forging employed, proper forging parameters can lead to a final product with very few defects and an excellent SME.

7.3.3.2 Rolling

Rolling is a manufacturing process in which the material is forced to pass between two inversely rotating rolls. The two rolls must rotate with the same absolute radial velocity in order to avoid high strains and defects. The opening between the two rolls is adjustable so that the rolling process can be performed in multiple passes. As the material is forced to pass through the opening of the two rolls, the material's thickness decreases, resulting in a subsequent increase in its width and mainly in its length. Commonly, the thickness reduction is performed under a logarithmic strain rate (Facchinello, Brailovski, Prokoshkin, Georges, & Dubinskiy, 2012). Similar to the other forming manufacturing processes, rolling can be employed either as a cold or hot process, with the same advantages and disadvantages that were mentioned before. The initial forms of the material can be ingots, slabs, or billets, according to the desired final form, which can be a plate, a sheet, a foil, or a ribbon. The use of lubrication is common in rolling, although it can be omitted.

Cold rolling is widely employed for SMAs. A lot of studies have been conducted which reveal that the results of cold rolling with a subsequent annealing in SMAs, produce fine microstructured or even nanostructured materials (Demers, Brailovski, Prokoshkin, & Inaekyan, 2009; Prokoshkin et al., 2008; Sharifi, Karimzadeh, & Kermanpur, 2014). Such nanostructured materials present an excellent SME, while their mechanical properties are improved.

The main factors that affect the structure formation during cold rolling are the thickness reduction during each pass, the magnitude pulling tension applied to the material, as well as whether lubrication is employed. The optimal thickness reduction for each SMA should be investigated before a cold rolling process. Further reduction of the thickness per pass results in increased work-hardening of the material, something that is undesirable. Lubrication during cold rolling helps maintain a low temperature as well as rendering the process more constant (Demers et al., 2009). In some cases, a thin layer of oxides may be intentionally left in the outer surface of the material, as it acts as a lubricant. On the other hand, if the oxide layer is thick, it can cause the formation of defects in the material's surface, as well as depress the material's SME and SE properties (Braz Fernandes, Mahesh, & dos Santos Paula, 2013). Finally, any pulling tension applied to the material reduces its springback effect, thus resulting in more corresponding dimensions in the product. On the other hand, applying excessive pulling tension causes the propagation of microcracks in the material's volume, rendering the material more susceptible to the formation of defects (Demers et al., 2009).

Hot rolling is also a commonly employed manufacturing process in SMAs. Overall, the factors that affect the quality of the product in this process are the same, with the addition of the applied manufacturing temperature. In this process, the factor of manufacturing temperature is crucial. If the rolling temperature is above a certain level, the material is submitted to dynamic recrystallization or to strain recovery. Both the recrystallization and the recovery, though, remain incomplete due to the fast cooling of the material in air (Braz Fernandes et al., 2013). In regard to the rest of the factors of the process, they also need to be properly selected, as they can heavily affect the hardness and work-hardening of the final product. Finally, after hot rolling, it is common to employ annealing in order to relieve some of the residual stresses induced due to the manufacturing process, although their magnitude is fairly low in comparison with the ones induced during cold rolling.

The only major difference between cold and hot rolling is the fact that in hot rolling the material can be manufactured more easily as the increased temperature is auxiliary to the material's ductility, rendering it easier for manufacturing. On the other hand, in hot rolling, some special equipment is required in order to heat and maintain the increased temperature of the material until the end of the process.

Researching the proper manufacturing factors is vital for rolling. Because of the large deformations occurring in this process, nonproper selection of these factors will result in a product with little or no SME properties. Particularly in the case of cold rolling, the manufacturing can be fairly difficult because of the rapid work-hardening of the material during the process (Wu, 2001).

7.3.3.3   Cold-drawing and extrusion

Cold-drawing is maybe the most common manufacturing process found in the industry of SMAs. In regards to the quality and state of the final product, cold-drawing presents many similarities to hot rolling, that was described above (Gall et al., 2008). During this process a SMA rod or wire is pulled through a die, resulting in an elongated form with a decreased outer diameter. Typical forms produced include wires, bars, rods, and tubes. It should be noted that in the case of tube production via this process, additional tools may be needed (e.g., nondeformable and deformable mandrels), in order for the product to have a fixed inner diameter (ASM, 1993).

In order for the material to be reduced from its bulk form to the desired dimensions, a number of passes with intermediate annealing processes is needed. The fact that most of the plastic deformations are large and they are imparted while the material is in the martensite phase, combined with the intermediate annealing processes, create texturing inside the material. This texturing heavily affects the properties of the material, compared to the material's properties in its as-cast form (Miller & Lagoudas, 2001). Similar to rolling, the passes performed for the outer diameter decrease are usual under a logarithmic strain rate per pass (Khaleghi, Khalil-Allafi, Abbasi-Chianeh, & Noori, 2013). The maximum reduction of a SMA in drawing is mostly defined by the chosen material, as well as the conditions of the process. The surface of a drawn product is fairly finished, while the final dimensions present little divergence from the desired values. In order to enhance these attributes, the use of lubricants or other protective layers is often employed during the cold-drawing process (Otubo, Mei, & Koshimizu, 1995). The average velocity of the process along with the maximum decrease of the outer diameter is mainly defined by the combination of the material used and the die properties. This is mainly due to the fact that different materials possess different ductility, thus some can be manufactured more easily than others. Furthermore, the die geometry and configuration define the final form of the extruded product, thus enabling more possibilities of cross-sections for the product.

Recently, there has been research presenting that hot-drawing is also feasible in SMAs (Jiang, Zhang, Zhao, Tang, & Yi, 2013; Yamauchi, Ohkata, Tsuchiya, & Miyazaki, 2011). Nonetheless, the majority of the drawn products are created via cold-drawing. Usually, after a number of passes, the cold-drawn product is subjected to annealing so that the residual tensile and compression residual stresses can be relieved, as well as to restore the material's SME properties (Stirling et al., 2011). Similar to the other processes, the annealing process in cold-drawn SMA products is usually performed under a protective environment or in a vacuum to prevent the formation of a thick oxide layer which would degrade the material's quality and its SME properties (Nam, Chung, Noh, & Lee, 2001; Wu et al., 1996). Moreover, depending on the material, proper annealing time and temperature are highly significant, as they can heavily affect the thermo-mechanical properties and the microstructure of the final product and thus its strain recovery properties (Frick et al., 2005; Khaleghi et al., 2013).

Extrusion is basically the same process as drawing. Their difference lies in the fact that, in extrusion, the force is applied through the bell of the die rather than its end. For this reason, the extrusion process is mainly used in materials with a larger diameter and shorter length. The extrusion force is applied to the material through a punch or a mandrel. Because of its mechanism, extrusion can lead to greater reductions in the material's outer diameter than cold-drawing. It should be noted that in the case of SMAs, extrusion is mainly used to produce thick-walled tubes or large diameter rods of certain length. After an extrusion, the product is usually subjected to annealing, similar to cold-drawing. Through the whole process, the use of lubricants is often required, leading to a product with improved surface finish.

As with the previous manufacturing process, attention should be paid to the proper manufacturing conditions, both in drawing and extrusion. One should thoroughly research the optimal conditions fitting the material used and the desired final form before proceeding to the actual process.

7.3.4 Machining

The term machining is used to describe processes that shape parts by removing unwanted material, which is carried away from the workpiece, usually in the form of a chip; evaporation or ablation may take place in some machining operations. The more narrow term cutting is used to describe the formation of a chip via the interaction of a tool in the form of a wedge with the surface of the workpiece given that there is a relative movement between them. These machining operations include turning, milling, and drilling, among others and are usually referred to as conventional machining processes. Abrasive processes such as grinding are also part of the cutting processes of great importance in contemporary industry. Other nonconventional machining operations that may or may not include physical contact between cutting tool and workpiece, or may not have a cutting tool in the same sense as conventional processes, or utilize thermal or chemical energy for removing material from the workpiece, are electrodischarge machining (EDM), laser machining, water jet machining, and electrochemical machining, just to name some.

Surveys indicate that a high percentage of the value of all mechanical components manufactured in the world comes from machining operations and that annual expenditure on machine tools and cutting tools are several billion euros for industrially developed countries (Childs, Maekawa, Obikawa, & Yamane, 2000; Trent & Wright, 2000). Manufacturing technology is driven by two very important factors, which are closely interconnected, namely, better quality and reduced cost. Modern industry strives for products with dimensional and form accuracy and low surface roughness at an acceptable cost while, from an economic point of view, machining cost reduction achieved through the increase of material removal rate and tool life without compromising surface integrity, especially for hard-to-machine materials like SMAs, is highly desirable. The unique properties of SMAs have drawn the attention of many researchers and a broad range of studies and commercial applications already exist. The current research trends and applications of SMAs include automotive, aerospace, and biomedical sectors while the areas of interest are constantly expanding. Applications refer mostly to actuators and implants, but there have been more than 10,000 SMA related patents in the United States alone and more than 20,000 worldwide, in various industrial areas (Jani, Leary, Subic, & Gibson, 2014). Most applications refer to the microworld regime for state-of-the-art products requiring accuracy, surface integrity, and complex shapes at an acceptable cost. Machining can provide all these characteristics and perform better compared to other manufacturing processes. However, there are limitations connected to materials and tool properties and thus it is imperative to further study the machining of SMAs.

As pointed out in previous paragraphs, although there are several SMA categories, machining literature pertains mainly to NiTi alloys. When working with hard-to-machine materials such as titanium and nickel alloys, obstacles pertaining to the machinability of these materials are imposed. The quality and cost goals are not easily achieved due to work-hardening, phase transformations, and, in the case of SMAs, due to phenomena closely connected to SMA characteristics. Low material removal rates and high tool wear are commonly observed. In the following paragraphs, conventional and nonconventional machining processes used in the case of SMAs are considered. Aspects of the processes (e.g., cutting tools and cutting conditions) are investigated and the findings of relevant researches are presented.

7.3.4.1 Conventional machining

Most of the work conducted on the conventional machining of SMAs pertains to NiTi or ternary alloys of NiTi, thus the information presented revolves around these materials unless stated otherwise. A first look at the machinability characteristics of titanium and nickel alloys can give an insight to the phenomena that take place when NiTi SMAs are considered.

Titanium's reactivity with the cutting tools, low heat conductivity, high strength at elevated temperatures, and low elastic modulus results in increased temperatures at the tool–chip interface, high dynamic loads, workpiece distortions, and rapid tool wear (Ezugwu, Bonney, & Yamane, 2003). Nickel-based alloys also present high strength and are considered hard-to-machine. Additionally, due to their austenitic matrix, nickel superalloys work harden rapidly during machining and tend to produce continuous chip which is difficult to control during machining (Choudhury & El-Baradie, 1998; Ezugwu, Wang, & Machado, 1999). The results of the above characteristics lead to accelerated flank wear, cratering, and notching, depending on the tool material and the cutting conditions applied. To avoid premature failure of the tool, low cutting speeds, proper tool materials, and cutting fluids are required (Ezugwu & Wang, 1997).

All the difficulties reported for titanium and nickel alloys separately apply for NiTi alloys as well. Furthermore, key features of SMAs such as pseudoelasticity, pseudoplasticity, and high ductility of NiTi alloys impose more difficulties when machining these alloys, leading not only to rapid tool failure but also to poor workpiece quality due to excessive burr formation, adhesions on the machined surface, and microstructure alterations of the workpiece material. Machining of NiTi is connected to large strains, high strain rates and temperatures on the workpiece surface and the layers underneath it, which in turn results in surface and subsurface defects such as the formation of a white layer and the development of microcracks (Ulutan & Ozel, 2011).

Turning of NiTi SMAs was the subject of the work presented by Weinert, Petzoldt, and Kötter (2004). Different cutting tools, namely indexable coated and uncoated cemented carbide, PCD, CBN, and ceramic inserts, were investigated. The research concluded that coated cemented carbide tools present reduced wear. More specifically, tools with eight alternating layers of TiCN and TiAlN were found to perform better even in comparison to tools with harder coatings like TiB_2; the latter exhibited cratering on the rake face due to tribo-chemical dissolving of the coating. Uncoated cemented carbide tools present extensive tool wear, ceramic cutting tools are not capable of machining NiTi alloys irrespective of the cutting parameters, PCD tools present notch wear which leads to sudden tool breakage, and CBN tool wear is higher in comparison to coated cemented carbide tools and, in combination with their high cost, they are not preferable. Regarding the cutting parameters, it was concluded that for the machining of NiTi with coated cemented carbide tools, higher cutting speeds than those recommended by the relevant literature and the tool providers can be used. Tool wear was reduced and surface quality was improved at a cutting speed of 100 m/min.

In another work by Weinert and Petzoldt (2004) on turning of NiTi, the significance of cutting fluids is discussed. High wear of the cutting tool leads to significant friction and thus higher temperatures in the cutting zone. Dry turning of NiTi results in chip burning and very high cutting forces; the use of a cutting fluid as lubricant/coolant prevents the chip from burning. Other investigations propose cryogenic cooling as being advantageous in comparison to dry and minimum quantity lubrication conditions regarding tool wear and surface quality (Kaynak, Karaca, Noebe, & Jawahir, 2013; Kaynak, Tobe, Noebe, Karaca, & Jawahir, 2014). Weinert and Petzoldt (2004) also discuss the effect of cutting fluids on chip breaking and the formation of burrs. Poor chip breaking in NiTi machining results in long continuous chip which in turn results in tool wear and is not affected by the use of cutting fluids; chip breaking is noticed only in low cutting speeds where tool wear is excessive. Burrs, as a result of the high ductility of NiTi alloys, are reduced when an emulsion is used in comparison to dry cutting. Burrs are also favored with small feed rates, which should also be avoided due to high tool wear.

Milling of a NiTi alloy used in biomedical applications (50.8 at% Ni–49.2 at% Ti) was studied by Guo, Klink, Fu, and Snyder (2013). The authors conducted quasi-static and split-Hopkinson pressure bar compression tests to evaluate the mechanical properties of the material. The tests showed high strength of the material under static and dynamic conditions indicating that it is more difficult to machine NiTi than Ti- or Ni-based superalloys. For milling, coated carbide inserts were used and once again a shorter tool life in comparison to milling conventional metals was observed. Increase of feed rate leads to increase in surface roughness, however, very small feed rate presents high surface roughness. By correlation, increase in the flank wear of the tool increases surface roughness. Furthermore, high ductility of NiTi is responsible for large exit burrs. Finally, from the subsurface microstructure and microhardness investigation it may be said that a smaller feed rate results in a thicker white layer, indicative of the phase transformation due to excessive loading and temperatures.

Weinert and Petzoldt (2004) investigated the drilling of NiTi tubes for the production of parts to be used in medical applications. They argue that the work-hardening in the subsurface zone is of importance when this machining operation is considered. Hardness of the material is increased with low cutting speeds or high feed rate. Furthermore, it is argued that the use of coated instead of uncoated cemented carbide tools exhibits no advantage in the case of drilling. Drilling was also investigated by Lin, Lin, and Chen (2000). The observation of the drilled surface morphology reveals numerous wavy tracks that are the result of the action of a blunt tool, adhesion and abrasive deformation which lead to vibration and damage of the twist drill. Drilling chips are continuous and have a yellowish color as a result of elevated temperatures and oxidation.

Some studies pertain to the production of microparts by micromilling (Piquard, D'Acunto, Laheurte, & Dudzinski, 2014; Weinert & Petzoldt, 2008). Micromilling was selected as complex geometries required in microparts to be used for microactuators and medical applications can be achieved. The studies investigate the optimal cutting conditions and the characteristics of SMA machining like burr formation in the microworld; a direct down-scaling is impossible because phenomena may differ in comparison to the processes discussed in the previous paragraphs. It turns out that micromachining is as difficult as conventional machining processes if not more so and the ranges for optimal cutting conditions are rather limited.

Generally speaking, when machining SMAs, the combined action of strain hardening and fatigue hardening causes a severe hardening effect and impairs the cutting rate. As a result, the workpiece quality is poor and tool wear unacceptably high, even for optimized cutting parameters and suitable cutting tools. These effects may also influence the shape memory characteristics. Researchers have turned to the study of nonconventional machining processes with the aim of improving machining of SMAs.

7.3.4.2   Nonconventional machining

This category of machining processes refers to mechanisms of material removal that include no contact between the tool and the workpiece. Using this process, tool wear is minimized or totally diminished but surface integrity may still be affected, mostly due to thermal loading. These processes are widely used in SMA machining, especially for components with very small dimensions.

Most nonconventional machining works pertain to EDM or wire electrodischarge machining (WEDM) of NiTi SMAs. The object of the studies is usually the surface and subsurface modifications that take place from the spark discharges during machining and the influence of various parameters on the material removal rate of the process. Studies on the influence of the machining conditions on the surface roughness (Theisen & Schuermann, 2004; Zinelis, 2007) indicate that with an increase of the working energy, surface roughness worsens; increase of working current, voltage, and pulse on time results in a thicker and more abnormal melting zone. Surface roughness also depends on the thermal properties of the workpiece material, namely, melting temperature and thermal conductivity (Chen, Hsieh, Lin, Lin, & Huang, 2007). When cutting conditions are such that they improve workpiece quality, changes in the subsurface of the workpiece material due to excessive heat generation may be observed (Manjaiah, Narendranath, & Basavarajappa, 2014). In a comparison between milling and EDM (Guo et al., 2013) authors found that EDM produced higher surface roughness than that of milling. Furthermore, a white layer was also observed that was less thick than that measured when milling was applied. It is argued by the authors that the white layer of EDM was produced by melting and rapid quenching while the white layer of milling was attributed to deformation-induced phase transformation, making the nature of each white layer fundamentally different. In Figure 7.6, the white layer underneath a milled and an electrodischarge machined surface are depicted. In Figure 7.6a, the white layer for two different feed speeds are given; increase in

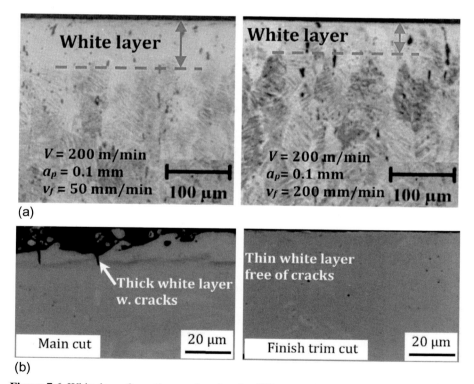

Figure 7.6 White layer formation produced under different cutting parameters for (a) milling and (b) EDM (Guo et al., 2013).

feed speed results in a thinner white layer. In Figure 7.6b, the white layer formed in an EDM processed workpiece with two different cutting parameters is observed. The white layer formed in EDM is thinner than that of milling. Furthermore, for finish trim cut, the white layer is even thinner and crack-free. Material removal rate also increases with increase in working energy (Theisen & Schuermann, 2004). Similar results are reported when WEDM is considered (Hsieh, Chen, Lin, Lin, & Chiou, 2009); surface roughness and material removal rate increase with an increase in machining energy. Lin, Lin, Chen, and Chu (2005) investigated the parameters of WEDM on Fe-based SMAs. They have measured a significant increase of the workpiece's hardness near the outer surface due to the formation of a recast layer. The recast layer is also responsible for a degradation of the shape recovery of the examined SMAs.

Laser machining of NiTi SMAs is also a promising process. As EDM, laser machining induces a heat affected zone on the workpiece material. NiTi alloys are sensitive to thermal influence and it is of interest to reduce the effect of excessive thermal loading when laser machining components are made of SMAs. A femtosecond laser is used for the machining of NiTi alloys for the production of microdevices (Li, Nikumb, & Wong, 2006). Although the ultrashort pulses of this laser perform better than other lasers, still the thermal nature of the process and the high ablation rates cause a significant recast layer. The authors suggest a sideways-movement path planning that permits better quality finished microproducts. The femtosecond laser machinability of NiTi was also investigated by Huang, Zheng, and Lim (2004). On the processed surfaces a redeposition layer and a heat affected layer were measured but were thinner than those observed from Nd:YAG laser machining, milling, and EDM while the resultant surface roughness was similar to that of precision milling. The laser drilling of ferromagnetic SMAs was investigated by Biffi and Tuissi (2014) and it was concluded that laser machining of microfeatures can be performed with limited thermal affection of the material.

Other works refer to abrasive water jet machining (Kong et al., 2013) and electrochemical polishing (Lee & Shin, 2011) of NiTi SMAs. Abrasive water jet machining for NiTi alloys exhibits no white layer but it was difficult to cut the material at straight kerf geometry. However, the authors' machining strategy indicated that this process can produce quality surfaces of NiTi SMAs. Works on nonconventional machining seem quite promising and are expected to further improve SMA products in the future.

References

ASM International. (1993). In *ASM metal handbook*: Vol. 14. *Forming and forging*. Materials Park, OH: ASM International.

Bellouard, Y. (2002). Microrobotics, microdevises based on shape-memory alloys. In M. Schwartz (Ed.), *Encyclopedia of smart materials* (pp. 620–644). New York, NY: John Wiley and Sons.

Biffi, C. A., & Tuissi, A. (2014). Fiber laser drilling of $Ni_{46}Mn_{27}Ga_{27}$ ferromagnetic shape memory alloy. *Optics & Laser Technology*, *63*, 1–7.

Braz Fernandes, F. M., Mahesh, K. K., & dos Santos Paula, A. (2013). Thermomechanical treatments for Ni–Ti alloys. In F. M. Braz Fernandes (Ed.), *Shape memory alloys—Processing, characterization and applications*. Available from: http://www.intechopen.com/books/ shape-memory-alloys-processing-characterization-and-applications/thermomechanical-treatments-for-ni-ti-alloys, Accessed 18.07.14.

Chen, S. L., Hsieh, S. F., Lin, H. C., Lin, M. H., & Huang, J. S. (2007). Electrical discharge machining of TiNiCr and TiNiZr ternary shape memory alloys. *Materials Science and Engineering A, 446*, 486–492.

Childs, T. H. C., Maekawa, K., Obikawa, T., & Yamane, Y. (2000). *Metal machining: Theory and applications*. Cambridge, MA: Elsevier.

Choudhury, I. A., & El-Baradie, M. A. (1998). Machinability of nickel-base super alloys: A general review. *Journal of Materials Processing Technology, 77*, 278–284.

Cong, D. Y., Wang, Y. D., Zhao, X., Zuo, L., Lin Peng, R., Zetterström, P., et al. (2006). Crystal structures and textures in the hot-forged Ni–Mn–Ga shape memory alloys. *Metallurgical and Materials Transactions A, 37*(5), 1397–1403.

Demers, V., Brailovski, V., Prokoshkin, S. D., & Inaekyan, K. E. (2009). Optimization of the cold rolling processing for continuous manufacturing of nanostructured Ti–Ni shape memory alloys. *Journal of Materials Processing Technology, 209*(6), 3096–3105.

Ezugwu, E. O., Bonney, J., & Yamane, Y. (2003). An overview of the machinability of aeroengine alloys. *Journal of Materials Processing Technology, 134*, 233–253.

Ezugwu, E. O., & Wang, Z. M. (1997). Titanium alloys and their machinability—A review. *Journal of Materials Processing Technology, 68*, 262–274.

Ezugwu, E. O., Wang, Z. M., & Machado, A. R. (1999). The machinability of nickel-based alloys: A review. *Journal of Materials Processing Technology, 86*, 1–16.

Facchinello, Y., Brailovski, V., Prokoshkin, S. D., Georges, T., & Dubinskiy, S. M. (2012). Manufacturing of nanostructured Ti–Ni shape memory alloys by means of cold/warm rolling and annealing thermal treatment. *Journal of Materials Processing Technology, 212*, 2294–2304.

Frick, C. P., Ortega, A. M., Tyber, J., Maksound, A.El.M., Maier, H. J., Liu, Y., et al. (2005). Thermal processing of polycrystalline NiTi shape memory alloys. *Materials Science and Engineering A, 405*(1–2), 34–49.

Funakubo, H. (1987). *Shape memory alloys*. New York, NY: Gordon and Breach Science Publishers.

Gall, K., Tyber, J., Wilkesanders, G., Robertson, S. W., Ritchie, R. O., & Maier, H. J. (2008). Effect of microstructure on the fatigue of hot-rolled and cold-drawn NiTi shape memory alloys. *Materials Science and Engineering A, 486*(1–2), 389–403.

Guo, Y., Klink, A., Fu, C., & Snyder, J. (2013). Machinability and surface integrity of Nitinol shape memory alloy. *CIRP Annals—Manufacturing Technology, 62*, 83–86.

Hsieh, S. F., Chen, S. L., Lin, H. C., Lin, M. H., & Chiou, S. Y. (2009). The machining characteristics and shape recovery ability of Ti–Ni–X(X=Zr, Cr) ternary shape memory alloys using the wire electro-discharge machining. *International Journal of Machine Tools & Manufacture, 49*, 509–514.

Hsieh, S., & Wu, S. (1998). Room-temperature phases observed in Ti53–xNi47Zrx high-temperature shape memory alloys. *Journal of Alloys and Compounds, 226*, 276–282.

Huang, W. M., Song, C. L., Fu, Y. Q., Wang, C. C., Zhao, Y., Purnawali, H., et al. (2013). Shaping tissue with shape memory materials. *Advanced Drug Delivery Reviews, 65*, 515–535.

Huang, H., Zheng, H. Y., & Lim, G. C. (2004). Femtosecond laser machining characteristics of Nitinol. *Applied Surface Science, 228*, 201–206.

Ito, K., Sahara, R., Farjami, S., Maruyama, T., & Kubo, H. (2006). Evolution of solidification structures in Fe–Mn–Si–Cr shape memory alloy in centrifugal casting. *Materials Transactions, 47*(6), 1584–1594.

Jani, J. M., Leary, M., Subic, A., & Gibson, M. A. (2014). A review of shape memory alloy research, applications and opportunities. *Materials and Design, 56*, 1078–1113.

Jiang, S.-Y., Zhang, Y.-Q., Zhao, Y.-N., Tang, M., & Yi, W.-L. (2013). Constitutive behavior of Ni–Ti shape memory alloy under hot compression. *Journal of Central South University, 20*(1), 24–29.

Kaynak, Y., Karaca, H. E., Noebe, R. D., & Jawahir, I. S. (2013). Tool-wear analysis in cryogenic machining of NiTi shape memory alloys: A comparison of tool-wear performance with dry and MQL machining. *Wear, 306*, 51–63.

Kaynak, Y., Tobe, H., Noebe, R. D., Karaca, H. E., & Jawahir, I. S. (2014). The effects of machining on the microstructure and transformation behavior of NiTi alloy. *Scripta Materialia, 74*, 60–63.

Khaleghi, F., Khalil-Allafi, J., Abbasi-Chianeh, V., & Noori, S. (2013). Effect of short-time annealing treatment on the superelastic behavior of cold drawn Ni-rich NiTi shape memory wires. *Journal of Alloys and Compounds, 554*, 32–38.

Kong, M. C., Srinivasu, D., Axinte, D., Voice, W., McGourlay, J., & Hon, B. (2013). On geometrical accuracy and integrity of surfaces in multi-mode abrasive waterjet machining of NiTi shape memory alloys. *CIRP Annals—Manufacturing Technology, 62*, 555–558.

Kumar, P. K., & Lagoudas, D. C. (2008). Introduction to shape memory alloys. In D. C. Lagoudas (Ed.), *Shape memory alloys, modeling and engineering applications* (pp. 1–51). College Station, TX: Springer-Verlag.

Lavernia, E. J., & Srivatsan, T. S. (2010). The rapid solidification processing of materials: Science, principles, technology, advances, and applications. *Journal of Materials Science, 45*(2), 287–325.

Leclercq, S., & Lexcellent, C. (1996). A general macroscopic description of the thermomechanical behavior of shape memory alloys. *Journal of the Mechanics and Physics of Solids, 44*(6), 953–957, 959–980.

Lee, E. S., & Shin, T. H. (2011). An evaluation of the machinability of Nitinol shape memory alloy by electrochemical polishing. *Journal of Mechanical Science and Technology, 25*(4), 963–969.

Li, C., Nikumb, S., & Wong, F. (2006). An optimal process of femtosecond laser cutting of NiTi shape memory alloy for fabrication of miniature devices. *Optics and Lasers in Engineering, 44*, 1078–1087.

Lidquist, P. G., & Wayman, C. M. (1990). Shape memory and transformation behavior of martensitic Ti–Pd–Ni and Ti–Pt–Ni alloys. In T. W. Duerig, K. N. Melton, D. Stöckel, & C. M. Wayman (Eds.), *Engineering aspects of shape memory alloys* (pp. 58–68). London: Butterworth-Heinemann.

Lin, H. C., Lin, K. M., & Chen, Y. C. (2000). A study on the machining characteristics of TiNi shape memory alloys. *Journal of Materials Processing Technology, 105*, 327–332.

Lin, H. C., Lin, K. M., Chen, Y. S., & Chu, C. L. (2005). The wire electro-discharge machining characteristics of Fe–30Mn–6Si and Fe–30Mn–6Si–5Cr shape memory alloys. *Journal of Materials Processing Technology, 161*, 435–439.

Liu, Y., Xie, Z., Van Humbeeck, J., & Delaey, L. (1998). Asymmetry of stress–strain curves under tension and compression for NiTi shape memory alloys. *Acta Materialia, 46*(12), 4325–4338.

Lojen, G., Anžel, I., Kneissl, A., Križman, A., Unterweger, E., Kosec, E., et al. (2005). Microstructure of rapidly solidified Cu–Al–Ni shape memory alloy ribbons. *Journal of Materials Processing Technology, 162–163*, 220–229.

Lojen, G., Gojić, M., & Anžel, I. (2013). Continuously cast Cu–Al–Ni shape memory alloy—Properties in as-cast condition. *Journal of Alloys and Compounds, 580*, 497–505.

Manjaiah, M., Narendranath, S., & Basavarajappa, S. (2014). Review on non-conventional machining of shape memory alloys. *Transactions of Nonferrous Metals Society of China, 24*, 12–21.

Miller, D. A., & Lagoudas, D. C. (2001). Influence of cold work and heat treatment on the shape memory effect and plastic strain development of NiTi. *Materials Science and Engineering A, 308*(1–2), 161–175.

Miyazaki, S., Kohiyama, Y., Otsuka, K., & Duerig, R. W. (1990). Effects of several factors on the ductility of the Ti–Ni alloy. *Materials Science Forum, 56–58*, 765–770.

Nam, T.-H., Chung, D.-W., Noh, J.-P., & Lee, H.-W. (2001). Phase transformation behavior and wire drawing properties of Ti–Ni–Mo shape memory alloys. *Journal of Materials Science, 36*(17), 4181–4188.

Olier, P., Barcelo, F., Bechade, J. L., Brachet, J. C., Lefevre, E., & Guenin, G. (1997). Effects of impurities content (oxygen, carbon, nitrogen) on microstructure and phase transformation temperatures of near equiatomic TiNi shape memory alloys. *Journal de Physique IV France, 7*(5), 143–148.

Otubo, J., Mei, P. R., & Koshimizu, S. (1995). Production and characterization of stainless steel based Fe–Cr–Ni–Mn–Si(–CO) shape memory alloys. *Journal de Physique IV France, 5*(8), 427–432.

Piquard, R., D'Acunto, A., Laheurte, P., & Dudzinski, D. (2014). Micro-end milling of NiTi biomedical alloys, burr formation and phase transformation. *Precision Engineering, 38*, 356–364.

Potapov, P., Shelyakov, A., Gulyaev, A., Svistunova, E., Matveeva, N., & Hodgson, D. (1997). Effect of Hf on the structure of Ni–Ti martensitic alloys. *Materials Letters, 32*(4), 247–250.

Prokoshkin, S. D., Brailovski, V., Inaekyan, K. E., Demers, V., Khmelevskaya, I. Y., Dobatkin, S. V., et al. (2008). Structure and properties of severely cold-rolled and annealed Ti–Ni shape memory alloys. *Materials Science and Engineering A, 481–482*, 114–118.

Sharifi, E. M., Karimzadeh, F., & Kermanpur, A. (2014). The effect of cold rolling and annealing on microstructure and tensile properties of the nanostructured $Ni_{50}Ti_{50}$ shape memory alloy. *Materials Science and Engineering A, 607*, 33–37.

Stirling, L., Yu, C.-H., Miller, J., Hawkes, E., Wood, R., Goldfield, E., et al. (2011). Applicability of shape memory alloy wire for an active, soft orthotic. *Journal of Materials Engineering and Performance, 20*(4–5), 658–662.

Strnadel, B., Ohashi, S., Ohtsuka, H., Ishihara, T., & Miyazaki, S. (1995). Cyclic stress–strain characteristics of Ti–Ni and Ti–Ni–Cu shape memory alloys. *Material Science and Engineering A, 202*, 148–156.

Theisen, W., & Schuermann, A. (2004). Electro discharge machining of nickel–titanium shape memory alloys. *Materials Science and Engineering A, 378*, 200–204.

Trent, E. M., & Wright, P. K. (2000). *Metal cutting*. Boston, MA: Butterworth-Heinemann.

Ulutan, D., & Ozel, T. (2011). Machining induced surface integrity in titanium and nickel alloys: A review. *International Journal of Machine Tools & Manufacture, 51*, 250–280.

Weinert, K., & Petzoldt, V. (2004). Machining of NiTi based shape memory alloys. *Materials Science and Engineering A, 378*, 180–184.

Weinert, K., & Petzoldt, V. (2008). Machining NiTi micro-parts by micro-milling. *Materials Science and Engineering A, 481–482*, 672–675.

Weinert, K., Petzoldt, V., & Kötter, D. (2004). Turning and drilling of NiTi shape memory alloys. *Annals of the CIRP, 53*(1), 65–68.

Wu, M. H. (2001). Fabrication of Nitinol materials and components. In: *Proceedings of the international conference on shape memory and superelastic technologies*, Kunming, China (pp. 285–292).

Wu, S. K., Lin, H. C., & Yen, Y. C. (1996). A study on the wire drawing of TiNi shape memory alloys. *Materials Science and Engineering A, 215*(1–2), 113–119.

Xu, Y., Otsuka, K., Toyama, N., Yoshida, H., Nagai, H., & Kishi, T. (2003). Additive nature of recovery strains in heavily cold-worked shape memory alloys. *Scripta Materialia, 48*(6), 803–808.

Yamauchi, K., Ohkata, I., Tsuchiya, K., & Miyazaki, S. (2011). *Shape memory and superelastic alloys: Applications and technologies.* Woodhead Publishing.

Yeom, J. -T., Kim, J. H., Hong, J. -K., Kim, S. W., Park, C. H., Nam, T. -H., et al. (2014). Hot forging design of as-cast NiTi shape memory alloy. *Materials Research Bulletin, 58*, 234–238.

Zhao, L. C., Duerig, T. W., Justi, S., Melton, K. N., Proft, J. L., Yu, W., et al. (1990). The study of niobium-rich precipitates in a Ni–Ti–Nb shape memory alloy. *Scripta Metallurgica and Materialia, 24*, 221–226.

Zinelis, S. (2007). Surface and elemental alterations of dental alloys induced by electro discharge machining (EDM). *Dental Materials, 23*, 601–607.

Index

Note: Page numbers followed by *f* indicate figures and *t* indicate tables.

Printed in the United States
By Bookmasters